At the Dawn of Humanity

At the Dawn of Humanity
The First Humans

BY
GERARD M. VERSCHUUREN

Angelico Press

First published in the USA
by Angelico Press 2020
Copyright © by Gerard M. Verschuuren 2020

For information, address:
Angelico Press, Ltd.
169 Monitor St.
Brooklyn, NY 11222
www.angelicopress.com

pb 978-1-62138-552-3
cloth 978-1-62138-553-0

Book and cover design
by Michael Schrauzer

TABLE OF CONTENTS

PREFACE

WHAT HAPPENED WHEN HUMANITY EMERGED? IT depends on whom you ask. Many scientists will say that almost nothing really changed. Many religious people will respond that almost everything was new. Most other people are somewhere in-between—between almost everything and almost nothing. The extreme ends of the spectrum will either minimize or maximize the differences between the non-human animal world and the human world. Who is right?

In this book we want to investigate first how genes may change from generation to generation—before, during, and after the dawn of humanity. After we have studied those mechanisms, we want to find out to what degree this can help explain what many people consider to be unique to humanity: the faculties of language, rationality, morality, self-awareness, and religion. Are those features really unique, or did they come from the non-human animal world? Were the first humans able to use language, think rationally, act morally, know who they were, and know there is a God? The answer may surprise you.

1

Common Descent

THE IDEA THAT WE ARE *PART* OF THE ANIMAL WORLD is much older than the idea that we came *forth* from the animal world. When Thomas Aquinas, for instance, followed Aristotle in calling a human being a "rational animal,"[1] he was certainly not talking evolution. He was merely categorizing human beings as animals, because they share all characteristics most animals have—to mention just a few, they all breed, feed, bleed, and excrete. But at the same time he distinguishes them as unusual and very peculiar animals who have the faculties of reason and intellect.

Aquinas did not need any learned biology to see the plain truth that humans are first of all animals. Obviously, Aquinas did not know about evolution and evolutionary biology—those concepts would be conceived some six centuries later—yet he could see humans as part of the animal world. He was in fact even able to acknowledge that "animals of new kinds arise occasionally from the connection of individuals belonging to different species, as the mule is the offspring of an ass and a mare."[2] This does not alter the fact, though, that Aquinas did not deal with issues such as common descent and natural selection—those concepts too would be conceived centuries later.

UNIFYING THE LIVING WORLD

Five centuries after Thomas Aquinas, Carl Linnaeus developed his famous Linnaean classification (taxonomy) of the living world, in which he classified organisms based on their similarities and dissimilarities—but again without thinking in terms of evolution. The similarities between organisms did not give him reason to assume they had common ancestors. In other words, Linnaeus did not speak in terms of "common descent." All he did was to base his scheme on the structural similarities of the different organisms. He actually thought that species were immutable,

1 Animal rationale or rationabile, e.g., in *Summa Theologiae*.
2 *Summa Theologiae* I, 73.1 ad 3.

although he deemed the creation of new species possible, albeit to a very limited extent. He even said, "We count as many species as there were created forms in the beginning."[3] It was only later in life that Linnaeus came across a new type of plant with a *hybrid* origin, which began to change his view. In one of his letters he wrote, "This new plant propagates itself by its own seed and is therefore a new species, not existing from the beginning of the world; it is a new genus never in being until now. It is a mule [i.e., hybrid] species in the vegetable kingdom."[4]

But soon things would be changing rather quickly. The observation that organisms show both similarities and dissimilarities was leading more and more people to the idea that their similarities could be explained by "common descent," and their dissimilarities by "descent with modification." Nowadays we would say that the idea of *evolution* was emerging. At one point (in 1973) the legendary biologist Theodosius Dobzhansky would word this even more poignantly, "Nothing in biology makes sense except in the light of evolution."[5]

However, the thought that all species have common ancestors, or common descent, was not really new. Several Greek philosophers had already adopted this idea, but the hypothesis got a real boost with William Harvey's claim in 1651 that all life emerges from life.[6] If this were so, then there wouldn't be any form of life without "parents." From then on, evolutionary ideas were set out by a few natural philosophers, including Pierre Maupertuis in 1745 and Erasmus Darwin in 1796. Around 1800, Jean-Baptiste Lamarck developed Lamarckism, the first coherent theory of evolution, proposing in *Philosophie Zoologique* (1809) and other works his theory of the transmutation of species and drawing a genealogical tree to show the biological connection of all organisms.

However, it was Charles Darwin who would give the idea of evolution a real boost with his concept of natural selection as a mechanism of evolution. He was still developing his theory in 1858 when Alfred Russell Wallace published a paper with a similar theory, so Darwin rushed to publish his

---

3 Carolus Linnaeus, *The Fundamenta Botanica*, No. 157 (1736).

4 In a letter to the Swiss naturalist Albrecht von Haller.

5 Theodosius Dobzhansky, "Nothing in Biology Makes Sense Except in the Light of Evolution," *American Biology Teacher*, 1973, 35 (3): 125–129.

6 Usually paraphrased as "omne vivum ex ovo [or vivo]." In *Exercitationes de generatione animalium* (London: 1651), Harvey actually says: "omnia omnino animalia, etiam vivipara, atque hominem adeo ipsum, ex ovo progigni."

book in 1859: *On the Origin of Species by Means of Natural Selection.*[7] In this groundbreaking work Charles Darwin launched his own evolutionary theory. Before he addressed the mechanism of evolution in his book, he documented the fact of evolution — the fact that there is "common descent." This essentially means that all living entities have a common origin in spite of apparent differences — so-called "common descent with modification." Interestingly enough, in his book Darwin never used the word "evolution" but always referred to "descent with modification."

Darwin elaborated extensively on various kinds of evidence for common descent in several areas of biology. He pointed out, for instance, that there is much morphological evidence for common descent among mammals. Such phenomena only make sense, in Darwin's view, if we assume that all mammals have common ancestors and inherited from them the same basic body plan. In his own words,

> The framework of bones being the same in the hand of a man, wing of a bat, fin of the porpoise, and leg of the horse — the same number of vertebrae forming the neck of the giraffe and of the elephant — and innumerable other such facts, at once explain themselves on the theory of descent with slow and slight successive modifications.[8]

NATURAL SELECTION

After citing much more evidence for "common descent," Darwin put forward his solution to one of the burning scientific questions of his day: What explains how there could be common descent "with modification"? How did organisms change and come to be as they are now? So not only did he argue for common descent, but he also suggested a causal mechanism for the process of modification by introducing the concept of *natural selection.* In his own words, "I am convinced that Natural Selection has been the main but not exclusive means of modification."[9] How natural selection is supposed to work we will discuss in the next chapter.

---

7 The full title is actually much longer, and rather revealing: *On the Origin of Species by Means of Natural Selection, or the Preservation of Favoured Races in the Struggle for Life.*

8 Charles Darwin, *Origin of Species* (London: John Murray), chapter 14, 479.

9 Ibid., Introduction, 6.

After the 1859 publication of Charles Darwin's *On the Origin of Species*, it soon became generally understood that classifications, such as the one made by Linnaeus, ought to reflect the phylogeny of organisms — their "family tree," so to speak. Since then, many types of so-called "evolutionary trees" have been constructed, showing the hypothetical connections between related fossils — their common descent as well as their modifications. But there is more to it.

The "modification" part not only explains how new features came along, but also how old features could be on their way "out." These latter features are usually called "rudimentary" (or "vestigial") structures in the context of common descent. Darwin explained them as structures that had lost their usefulness due to changes in lifestyle and environment. He speaks of "use and disuse."[10] We can find them all over the mammal world. Moles still have eyes, but they are underdeveloped. Whales still have a tiny, underdeveloped pelvis dating back to when they were land mammals. We ourselves share ear muscles and an appendix with other mammals, although these have lost their original function. As a matter of fact, all mammals still have floating ribs and vertebral processes that were most likely real ribs in their long-gone ancestors. And in the evolutionary tree of horses, we observe that certain parts of their hoofs gradually disappear while other parts have become very prominent — by "use and disuse." Rudimentary features are anomalies in a world of fixed species; but in a world of common descent they do make sense. A trait that was formerly beneficial — such as the wings of ostriches and penguins that do not fly, or the webbed feet of upland ducks that rarely go near the water — may still be inherited even though no longer beneficial.

But there is also a serious problem with evolutionary trees, namely the fact that certain connections between organisms seem to be uncertain or even missing — the so-called "missing links." In evolutionary trees we must assume some kind of link between fishes and reptiles, reptiles and birds, or reptiles and mammals. Darwin explained these missing links by assuming an incomplete geological record: "The explanation lies, as I believe, in the extreme imperfection of the geological

10   Ibid., chapter 5, 135.

record."[11] Indeed, fossilization is a relatively rare phenomenon, therefore many connections may be missing, which would explain those so-called "missing links."

When looking at the fossil record we find it to be full of gaps—forms appear and disappear in a seemingly sudden way. There are hardly any transitional forms. After Darwin, many have tried to "unearth" missing links to fill the gaps between fishes and amphibians (*Ichthyostega*), between amphibians and reptiles (*Seymouria*), and between reptiles and birds (*Archaeopterix*), to name just a few. They are portrayed to be fossils of transitory stages, such as fish-like reptiles, reptile-like birds, and reptile-like mammals. But such cases were hotly debated and are still quite controversial.

On the other hand, we can still find existing intermediate forms between fishes and reptiles (such as amphibians, which grow up in water like fish, move onto land like reptiles, and come back to water to lay their eggs), or between reptiles and mammals (think of so-called living fossils such as a duck-billed, beaver-tailed, otter-footed mammal in Australia, called *Platypus*, known to lay eggs developed in a "mammalian" uterus). In other words, the particular shape of evolutionary trees may be up for discussion, calling for some educated guess work, but the tree itself is basically accepted by most biologists. Fortunately, nowadays we have newer and better tools at our disposal to assess evolutionary connections in more reliable and more scientific ways, as we will discuss later (see chapter 2).

So, we have to revisit the question whether the idea of "common descent" is very credible. Undoubtedly, its appealing part is that it creates unity in the midst of diversity. The living world is characterized by an enormous diversity—from plants to animals, from insects to mammals, from mice to elephants—and yet there is a striking unity: they all share many commonalities and seem part of one gigantic family tree. Accepting this idea would make much more sense of the way we ourselves fit into a classification of the animal world. If this is true, then humans too seem to have many "relatives" in the animal world, some looking like close relatives and others like more distant relatives. In fact, humans have a skeleton like that of some of the animals, the

11 Ibid., chapter 9, 346.

vertebrates. In addition, they start their lives in a womb, just as some of the vertebrates, the mammals, do. Then again, they have a relatively large brain, something they have in common with a subgroup of the mammals, the primates. And like some of these primates, the hominids, they lack a tail. When tightening the range, the resemblances become more and more numerous and striking. This makes it easier for us to see that we are all part of the same family tree that connects all that lives on Planet Earth.

2

Evolution Step by Step

ONCE DARWIN HAD DISCUSSED CONSIDERABLE EVI-
dence in favor of common descent, he went into the issue of the mechanism behind the modifications in common descent. It was his conviction,
as he said in one of his letters, that evolution follows laws in the same
way as planets and comets follow laws in the physical sciences. In his
own words, "astronomers do not state that God directs the course of each
comet and planet."[1] Darwin compares what he did for the evolution of
species to what Newton had done for the orbits of planets.[2] Well, one
of the main laws Darwin came up with is the "law" of natural selection.

SURVIVAL OF THE FITTEST

Darwin always felt uneasy about the term "natural selection." He did
not like the idea that this term leads almost automatically to the obvious question of "selection by whom or by what?" The implication of a
"selecting agent" looms large. Darwin certainly tried to avoid this implication by saying he had as much right to use metaphorical language as
physicists do. In his own words, "who objects to an author speaking of
the attraction of gravity as ruling the movements of the planets? Everyone
knows what is meant and is implied by such metaphorical expressions."[3]
Yet, his colleague Alfred Wallace convinced Darwin to finally replace the
term "natural selection" with Spencer's notion of "survival of the fittest"
in the 5th edition of his book.

Darwin stirred quite some controversy with his new theory—not so
much for biological reasons as for theological reasons. He seemed to
replace the species God had created with the species nature produces. Darwin defended his approach with the distinction between "secondary causes"
in evolution and "a primary cause," the Creator. This allowed him to say,

---

1 In correspondence with geologist Charles Lyell in 1861.
2 In his own words, "[W]hilst this planet has gone cycling on according to the
fixed law of gravity, from so simple a beginning endless forms most beautiful and most
wonderful have been, and are being, evolved" (ibid., chapter 14, 490).
3 Ibid. (5th edition), chapter 4, 93.

"To my mind it accords better with what we know of the laws impressed on matter by the Creator, that the production and extinction of the past and present inhabitants of the world should have been due to secondary causes, like those determining the birth and death of the individual."[4]

Interestingly enough, Darwin adopted a metaphysical principle that had already been formulated by St. Thomas Aquinas. Seen in this context, it can be stated that God as the Primary Cause operates in and through secondary causes, including the causes of evolution. Creation is about God, the Primary Cause, whereas evolution would be about secondary causes such as mutation and natural selection. Creation creates something out of nothing, whereas evolution produces something out of something else—by following biological laws in the same way as planetary motions follow physical laws. God does not make things himself, but he makes sure they are being made through his laws of nature. Had Thomas Aquinas known about evolution, he would probably have said that evolution offers us a scientific account of changes, of how a later state of the material world might have emerged from an earlier state—whereas Creation offers us a metaphysical account of where the material world itself ultimately comes from. Darwin somehow followed this classic distinction.

A more serious attack on Darwin's theory is the accusation that even the concept of "survival of the fittest" is problematic because it amounts to a tautology—that is, a statement that is true in every possible interpretation. The argument goes like this: "Who survives? The fittest! Who are the fittest? Those who survive!" I don't think, though, we need take this attack seriously, for it is based on some terminological confusion. Biological fitness has actually a double meaning: it refers to the role of an organism as the subject (or cause) of reproduction, but also to the role of an organism as the object (or effect) of reproduction. This ambiguity also affects the fitness concept: Darwin's concept of fitness—let's call it D-fitness—is potential reproductive success (a cause), but there is also the concept of fitness in the sense of actual reproductive success (an effect), which is the way the statistician Robert Fisher and some population geneticists use it—let's call it F-fitness. Apparently, the slogan "survival of the fittest" is not a tautology if we take fitness as D-fitness, or reproductive capacity. The principle of natural selection (or survival of the fittest) asserts that those organisms that are potentially successful in reproduction are more likely to be also actually successful in reproduction. Having a good "design" does matter in evolution. Good designs are successful designs.

Another source of confusion about natural selection is that it promotes optimal designs, but not the perfect designs that some claim. Darwin himself may bear some of the blame for this. According to him, natural selection carries the evolutionary process to ever-greater perfection. In the *Origin of Species* he uses the word "perfect" 77 times, "perfected" 19 times, and "perfection" 27 times. However, perfection is not taken here in its absolute sense — it is, rather, relative perfection, that is, as best as it can be in order to accomplish what it is supposed to accomplish for the time being. Therefore, good designs can actually become outdated designs in Darwin's theory — designs that lost their functionality and thus became "rudimentary." The reason is simple: natural selection works under constantly changing circumstances. For instance, a feature such as the pelvis of whales lost its function when they moved to the ocean; and on land, the swim-bladders of fish may have changed into lungs. So the designs promoted by natural selection are optimal, not perfect. All that matters for an optimal design is reproductive success. Put in a slogan: Success breeds success! Of course, organisms do not calculate costs and benefits, but natural selection makes them act as if they do make strategic decisions.

There may be several reasons why an "optimal" design is usually not a "perfect" design. First of all, when balancing the benefits against the costs, there are compromises to be made. For instance, in humans, the long and flat pelvis of quadrupeds has become more basin-shaped with a short, broad top bone, good for balance but bad for birthing because it narrows the birth canal. Second, organisms live with built-in boundaries caused by their ancestral history and many other genetic and developmental constraints — which fact puts each specific design in a straitjacket so to speak. Third, there are constraints found in the "makeup" of this Universe, bound by laws of nature. Legs are better for locomotion than wheels because it is physically difficult to provide control and blood supply to a rotating object. Fourth, we need to stress that there is more to evolution than natural selection. Even Darwin admitted this; he called selection "the main but not the exclusive cause" of evolution.[5] There is also the role of random, stochastic processes. Sometimes "survival of the fittest" boils down to "survival of the luckiest." This random effect becomes even more pronounced when only small parts of a population migrate.

As a consequence of all this, designs are often far from perfect, but they are the best we can get in this Universe for the time being and given the time past. Natural selection goes for the best available design under the given circumstances; as a result, designs good for life in the Sahara may not be good for life in Alaska, and designs good for living in the trees may not be good for living on the ground, and so on. Due to genetic and hereditary differences caused by mutations, organisms differ in design, which allows natural selection to treat them "selectively." It leads to selective reproduction. In Darwin's own words, "This preservation of favourable variations and the rejection of injurious variations, I call Natural Selection."[6]

The effect of natural selection can be illustrated nicely with examples based on camouflage. Selective reproduction can explain why caterpillars, who feed on green cabbage, have a green color. Those among them who have a green color have a selective advantage over those of a different color; they have a better chance of eluding predators (D-fitness), and thus a better chance of producing more progeny, which leads to an increase of caterpillars with a green color in the next generations (F-fitness). There is a similar story about the white color of polar bears. At one time the Anglican Bishop Hugh Montefiore (1920–2005) tried to prove that natural selection couldn't have favored the successful design of a white camouflage among polar bears, because bears don't have predators, he said.[7] What he forgot, though, is that camouflage works in two directions—for predators as well as prey. For a polar bear with dark fur, it would be much harder to sneak up on a seal.

These are just very simple examples of how natural selection works. But a much more serious consideration needs to be addressed also. Darwin realized that natural selection on its own is very limited. It does not work if there is not some degree of pre-existing diversity in the population; if all members were identical, selection would not make any difference. He

---

5   *On the Origin of Species* (London: Murray, 1859), 6.
6   Ibid., chapter 4, 81.
7   Hugh Montefiore, *Probability of God* (London: SCM Press, 1985).

also realized that natural selection cannot generate diversity—it can only select from what already exists. Natural selection requires diversity but cannot spawn it. But as to how diversity could come about, Darwin did not really know at the time; the field of genetics had not yet emerged. With some hesitation, he adopted a form of blending heredity, although he realized that this leads to intermediate forms, so ultimately any favorable trait that might arise in a lineage would have "blended away" long before natural selection had a chance to work.

This problem remained until it would be solved a few decades after Darwin, when genetics began to flourish. This development gave a new boost to the Darwinian Theory, making for the rise of a new Darwinian theory—sometimes called Neo-Darwinism, or the Modern Synthesis. It integrates Darwin's ideas with new developments in genetics.

GENETICS

The field of genetics studies how genes affect the appearance of organisms and how they are transmitted to the next generation. A model very popular in population genetics is the gene-pool model: It represents the set of all genes in a population. The simplest models of a gene pool focus on one specific gene with its variants, called alleles. The gene-pool model simulates how selection pressure on this particular gene may change allele frequencies by selectively promoting certain alleles of this gene over others. To put it briefly, it simulates the "battle" of alleles based on one particular gene.

To get a bit more technical: If the gene pool is composed of only two alleles, say A and a, then the frequency of allele A and the frequency of allele a will stay the same generation after generation, unless natural selection comes into play. The allele frequencies would remain the same, generation after generation, if there were no natural selection. But at the moment natural selection kicks in, alleles may have different probabilities of being transmitted to the next generation. In other words, they differ in D-fitness relative to each other: Some have a selective advantage, others a selective disadvantage.

So now the question is: How does this diversity or variability come about? Research has shown that most populations—gene pools, if you will—harbor a tremendous amount of variability. How is that possible? Where does this variability come from? The main factor behind variability is *mutation*—a change in genes, more particularly, in their DNA. Each organism has two copies of all its genes—one came from the father, the other from the mother—but they may not be complete copies of each other due to mutations.

Sickle-cell anemia, for instance, is caused by a single mutation in the DNA that produces hemoglobin. The mutation may cause a change in one of the two alleles for the hemoglobin gene. Hemoglobin in the blood carries oxygen from the lungs to the rest of the body. Now, a mutation may cause a change in one allele of those two copies—let's say, changing it from allele *H* into allele *h*. So the individual who had two of the same alleles, *HH*, now has two different alleles, *Hh*. This in itself does not cause anemia, because the non-mutated allele *H* still produces functional hemoglobin. However, it is possible that a child receives the *h*-allele from both parents, which would cause sickle-cell anemia. (This is a simplified explanation.)

In this view, mutations are the driving force behind genetic variability, on which natural selection depends. They create the variability and diversity in the genetic material from which natural selection can "select." The bottom-line is that natural selection in itself is not creative; it only selects, and thus favors or rejects what is already there.

Now let's bring things together in one simple example: DDT-resistant mosquitoes make an enzymatic protein that can accommodate a single molecule of DDT and inactivate it by adding oxygen to a chlorinated side group on its molecule.[8] Every once in a while a malaria mosquito has a gene mutation (mutating from allele *D* to allele *d*) that is capable of inactivating DDT. From now on the population of mosquitoes has some genetic diversity in this respect. Natural selection will probably not have much impact until the population has to deal with DDT. From then on the ones with the mutant DNA code have a selective advantage and will transfer their code much more frequently to the next generation than the carriers of the initial code. As a result, the genetic makeup of the population dramatically changes due to the pressure of natural selection. In time, nearly the entire population may become resistant.

But there is a serious problem with mutations: They are considered

8   J.M. Riveron and C.S. Wondji, "A single mutation in the GSTe2 gene allows tracking of metabolically-based insecticide resistance in a major malaria vector," *Genome Biology*, 2014.

random. Although randomness is a statistical, stochastic concept, in biology it can mean several more specific things. First of all, mutations are considered random in the sense of "spontaneous"—they just pop up. We know that certain factors may induce mutations, but we don't know when they occur. Second, mutations are—as far as we know—"unpredictable" as to where they strike. We cannot predict at what location in the DNA mutations will hit and what changes they might generate there. Third, mutations are random in the sense of "arbitrary," because mutations do not select their target but hit indiscriminately—"good and bad" spots alike, so to speak, for they have no "preferences." Fourth, mutations are also random in the sense of "aimless," because they occur without any connection to immediate or future needs of the organism. In that specific sense they can be considered "short-sighted." The reason is that there is no physical mechanism that detects which mutations would be beneficial and then causes those mutations to occur, so they lack any "fore-sight." In short, whereas natural selection has a definite "preference" for what fits best—for "the fittest," that is—mutations do not.

So the question is now: if mutations are random, how could something new, let alone something positive, come out of evolution? If mutations are indeed random—in the sense of unpredictable, arbitrary, and short-sighted—how "creative" can they be? Can they ever be advantageous for the organism and thus help evolution? It is true, mutations are mostly associated with defects and diseases such as albinism, dwarfism, phenylketonuria, color-blindness, sickle-cell anemia, etc., because a change in a "balanced," time-tested gene is more likely to be disruptive than constructive.

Most mutations occur at the gene level, by changing the sequence of nucleotides in DNA. Some mutations may not even cause much of a difference. The basic explanation is simple: the building blocks of DNA are nucleotides, which come in 4 different forms (A, C, G, and T), whereas the building units of proteins are amino acids, which come in 20 different shapes. The situation is similar to the numbers and letters on a phone. There are 10 different numbers but 26 different letters, so a specific number may stand for three different letters (e.g., 2 represents ABC). Because the coding unit of DNA, a *codon*, is 3 nucleotides long, which makes for 64 possible combinations ($4^3 = 4 \times 4 \times 4$), there is room for "synonyms" (e.g., the codons GCA, GCC, GCG, and GCU all specify the same amino acid, alanine). So a mutation from GCC to GCG would have no effect whatsoever.

However, more often than not, mutations do have some effect. They may change something that had worked fine so far, so the effect may vary from minor to major. When they occur in the gene's coding sequence for proteins, they may alter the functionality or stability of its protein product. Chances are they disrupt the code that had worked fine before. Nevertheless, there are also mutations that can be beneficial to the organism. The following list has just a few examples of various mutations that have been advantageous for the human gene pool.

1. All humans have a gene for a protein called apolipo-protein AI, which is part of the system that transports cholesterol through the bloodstream. A small community in Italy is known to have a mutated version of this protein, which is even more effective than Apo-AI at removing cholesterol from cells and dissolving arterial plaques.[9] Perhaps something similar is the case among the Maasai of Kenya who are heavy meat eaters and milk consumers, yet they have generally low cholesterol levels.

2. One of the genes that govern bone density in human beings is called LRP5. Mutations which impair the function of LRP5 are known to cause osteoporosis. But a different kind of mutation can amplify its function, causing one of the most unusual human mutations known.[10] This mutation was first discovered accidentally, when a young person from a Midwest USA family was in a serious car crash from which he walked away with no broken bones.[11]

3. Many adults lack a beneficial mutation that enables others to keep eating milk and dairy products after growing up—it is called lactose intolerance and is common in many populations in Eastern and Southeastern Asia, but also among some Africans.[12] The beneficial mutation is caused by a gene called LCT that allows the enzyme lactase to remain

9   G. Franceschini, et al., "Relation between the HDL apoproteins and A-I isoproteins in subjects with the AIMilano abnormality," *Metab. Clin. Exp.* 1981, 30 (5): 502–9.

10   Frost M, et al., "Patients with high-bone-mass phenotype owing to Lrp5-T253I mutation have low plasma levels of serotonin," *Journal of Bone and Mineral Research*, 2010, 25 (3): 673–75.

11   Y. Gong, a.o., "LDL Receptor-Related Protein 5 (LRP5) Affects Bone Accrual and Eye Development," *Cell*, vol. 107, 513–23, 2001.

12   D.M. Swallow, "Genetics of lactase persistence and lactose intolerance," *Annual Review of Genetics*, 2003, 37: 197–219.

expressed in infants after the mother has stopped breast feeding. Within the past 10,000 years, the mutation for this gene became beneficial and had a positive fitness value among those who practiced dairy farming.

4. There is also a mutation in a membrane receptor protein that either confers resistance to HIV or delays AIDS onset by preventing HIV from binding to cells.[13] It is most likely a remnant from resistance to the bubonic plague or smallpox. This might explain why this mutation is not found in southern Africa, where the bubonic plague never reached. Because HIV has not been around for more than one or two generations, there presumably has not been long enough selective pressure for it to spread throughout the population.

5. EPAS1 is a gene that codes for a protein involved in responding to a falling oxygen level. It seems to be the key to Tibetan adaptation to life at high altitude where there is 40% less oxygen in the air than at sea level.[14] A mutation in the gene that is thought to affect red blood cell production is present in only 9% of the Han population, but was found in 87% of the Tibetan population.

GENE DUPLICATES

In addition to mutations at the gene level, there are also mutations above the gene level, which can create *duplicates* of a certain section of the DNA code. Duplicates are rather common in anyone's entire DNA set — that is, in anyone's genome. To estimate how many duplicates of various genes a genome carries, geneticists use DNA micro-array technology. By using this method, it was found, for instance, that about 15% of the genes in the human genome appear to have been produced by gene duplication.

The duplication of a gene results in an additional copy that is free from selective pressure — that is, mutations of it have no direct deleterious effects for its host organism. Thus the duplicated gene can accumulate mutations faster than a functional single-copy gene, over generations of organisms. This freedom from consequences allows for the mutation of novel genes that could potentially code for a new function and increase

---

13   A.P. Galvani and J. Novembre, "The evolutionary history of the CCR5-Delta32 HIV-resistance mutation," *Microbes and Infection / Institut Pasteur*, 2005, 7 (2): 302–9.

14   C. Jeong, et al., "Admixture facilitates genetic adaptations to high altitude in Tibet," *Nature Communications*, 2014, 5: 3281.

the fitness of the organism. If the mutation in the duplicate is harmful, the duplicate becomes a so-called *pseudo*-gene. If the mutation is beneficial, then we may have a so-called *proto*-gene — "a gene-in-the-making."

The existence of duplicates is most likely also the explanation of how the human blood clotting mechanism came along. Blood clotting follows an intricate cascading route with a dozen or so proteins; these proteins are often called "factors." Most hemophiliacs carry a mutated version of factor VIII, some of factor IX. Most of these proteins turn out to be related to one another at the level of amino-acid sequence, which is an indication of ancient gene duplications, most likely in "silent" DNA sections that underwent mutational changes without having any direct impact on survival.

This explanation is even more plausible when we realize that fish have a much shorter cascade of blood-clotting proteins, while the longer cascade found in mammals has extra proteins very similar to the original ones, which makes it highly probable again they are a result of replicated DNA sections. Since new copies were not essential for the original function, they could gradually evolve to take on a new function, driven by the force of mutation and natural selection. Our current blood clotting mechanism is able to stop possible leaks much more quickly as it evolved from a low-pressure to a high-pressure cardio-vascular system — which required extra new proteins.[15]

In short, there are indeed mutations that can be beneficial for the organism, or at least potentially so, and can therefore be promoted by the mechanism of natural selection. So let's go back to the role of natural selection in evolution. Darwin's theory of natural selection is basically so simple that one might wonder why it took so long to emerge. But not only that. It is so simple that one might also wonder whether it could explain everything that happened in evolution. Can it?

GRADUALISM

There is nothing in Darwin's work that he repeats as often as the view that nature does not make leaps (*"natura non facit saltus,"* or *"saltum"* for the singular version). Darwin's contemporary Thomas Henry

---

15 R. F. Doolittle, "The evolution of vertebrate blood coagulation: A case of yin and yang," *Thrombosis and Haemostasis*, 1993, 70: 24–28.

Huxley—known as "Darwin's Bulldog" for his advocacy of Darwin's theory of evolution—warned Darwin that he had loaded his work "with an unnecessary difficulty in adopting *Natura non facit saltus* so unreservedly."[16] Yet this principle has given many biologists—not only then, but now as well, or even more so—the conviction that evolution is always gradual and slow, with short steps and small mutations, one small step at a time. In other words, Darwin's theory searches for small causes that may lead to large effects—one by one, step by step, gene by gene, generation after generation, species after species. Here we have the core of Darwin's view: varieties are the evolutionary basis for species; species are the evolutionary basis for larger units such as genera—and so on upwards in the hierarchy of taxonomy. What, ultimately, it all boils down to are small gene changes with large effects.

Many people have their doubts whether mutation-by-mutation steps can explain much in evolution, especially so when it comes to very complicated biological structures such as eyes or brains. Some biologists even have their doubts whether Neo-Darwinism can achieve what Darwin thought he could achieve. They speak of cases of "irreducible complexity,"[17] and argue that such complex structures can only work when all the components are in place, which does not seem possible with a step-by-step, gene-by-gene, mutation-by-mutation approach. They would raise objections like "What good is half an eye?"

Neo-Darwinians would respond, "half an eye is better than no eye at all." They use the following analogy to make their case: once the keystone is placed in an arched stone bridge, the scaffolding can be removed; from then on, removal of any part may cause the bridge to collapse, but that doesn't mean the bridge wasn't the product of a gradual construction process.[18]

Let's use again the example of the intricate cascading route of human blood-clotting, which involves many proteins. We find a partial clotting cascade in fish: some of these proteins have a long history in the animal

16 Thomas Henry Huxley, *Letter to Charles Darwin*, Nov. 23, 1859.

17 Michael Behe, *Darwin's Black Box* (New York: Free Press, 1998), 39.

18 E.g., Niall Shanks, *God, The Devil, and Darwin: A Critique of Intelligent Design Theory* (Oxford: Oxford University Press, 2004), 185. Or A.G. Cairns-Smith, *Seven clues to the origin of life: a scientific detective story* (Cambridge: Cambridge University Press, 1986), 61.

world, making for a partial cascade—far better than none. But when required to stop possible leaks much more quickly, our current blood clotting mechanism had to evolve from a low-pressure into a high-pressure cardio-vascular system—which called for a longer series of more diverse proteins. Since most of these proteins turn out to be related to one another at the level of amino acid sequence, they most likely reflect ancient gene duplications. Since new copies were not essential for the original function, they could gradually evolve to take on a new function, driven by the force of mutation and natural selection.

So the contrary assumption that the entire cascade had to emerge fully functional from the very outset as a complete set of DNA sequences seems to be unwarranted. Gradual evolution—with short steps and small mutations, one step at a time—seems to be possible in at least a series of cases. Whether there is more to it, we will find out soon.

3

Speciation Step by Step

ALTHOUGH MANY PEOPLE MAY ACCEPT THE IDEA
that genetic diversity (based on mutation) combined with natural selection
can produce changes in a *population*—as we see it happen, for instance,
in our domestic dogs and cats—the question remains: can this ever lead
to a new *species*? So the real problem arises when the theory of natural
selection tries to deal with evolution beyond the species boundaries: How
does a new species evolve? Is natural selection capable of achieving this?

Let's face it, no matter how long we keep applying artificial selection to
our enormous variety of dogs or cats, all individuals seem to remain part
of the same species and can still interbreed with other members of their
species. All dogs, no matter how different in appearance, still belong to
the species *Canis lupus*, and all cats to the species *Felis catus*. Critics of
the theory of natural selection keep hammering on this point; they stress
that the fruit fly *Drosophila melanogaster*, for instance, may undergo very
noticeable, dramatic mutations—often artificially enforced by extensive
radiation in the lab—but those mutated flies remain members of the
same species because they can still mate with other members of their
species. So how could a new species ever arise by means of mutation
and natural selection alone?

THE SPECIES CONCEPT
As a matter of fact, the claim that new species do arise is very credible.
The fossil record tells us that most species didn't make it in evolution;
they either became extinct or evolved into other species—but we have
never seen it happen. It's estimated that 99.9% of all species that ever
lived no longer exist. But what about a new species? The evolutionary
process by which a biological population supposedly evolves into a distinct, new species is called *speciation*. Is there such a process, and if so,
can it be achieved by mutation and natural selection alone?

Before we can answer questions like these, we need to find out
what a species is. It is very tempting to consider a species a group of
similar-looking organisms. However, that is a very hazy conception: How

similar is "similar," and where should we draw the line? Some people would rather adopt Ludwig Wittgenstein's concept of "family resemblances."[1] Wittgenstein noted that members of a family have overlapping similarities, even if they don't have one essential feature in common. Nevertheless, the idea of overlapping similarities is still too ambiguous to be useful in biology.

Besides, this approach of searching for similarities has been condemned by many biologists as a form of Aristotelian essentialism—a theory that seeks to categorize living things by positing the existence of shared natures, defined by a set of shared properties. However, it should be noted that this criticism misses the point. Aristotle's essentialism is not an attempt to place biological organisms into taxonomic categories. It is not about species membership. Rather it is a metaphysical theory that seeks to explain why organisms are what they are.[2]

Nevertheless, most biologists felt compelled to adopt a much more precise species concept—the so-called biological or reproductive species concept, which has proven to be very successful for population biologists. It was particularly developed and advocated by biologists such as Theodosius Dobzhansky from Columbia University, Francisco Ayala from the University of California at Irvine, and Ernst Mayr from Harvard University. It has been phrased in the following terms: Species are groups of interbreeding natural populations that are reproductively isolated from other such groups.[3]

According to this species concept, a species is an evolutionary unit that is, on the one hand, kept together by internal gene flow and, on the other, isolated from other species by intrinsic isolating mechanisms of reproduction. Its members may share some similarities, but what is more important is the fact that they are isolated from other groups by reproductive barriers. So this species concept is all about "family ties," and not so much "family resemblances." Similarities as such do not count for much, as they are believed to be merely a by-product of reproductive isolation.

---

1 Ludwig Wittgenstein, *Philosophical Investigations* (Oxford: Blackwell Publishing, 2001 [1953]), §67.

2 E.g., Denis Walsh, "Evolutionary Essentialism," *British Journal for the Philosophy of Science*, 57 (2006): 425–48.

3 See, e.g., E. Mayr, "Speciation phenomena in birds," *American Naturalist*, 1940, 74: 249–78.

The outcome of this discussion is that a species is no longer defined by properties that are based on shared characteristics, but by relationships, based on shared reproduction and reproductive isolation from other species. But if that is right, then Darwinism could have a problem with its mechanism of natural selection, because all cases of natural selection seem to be located within the boundaries of the species. If it's true that a species is characterized by reproductive isolation, we should ask ourselves how on earth a new species could ever have another species as its "parent." In other words, how can a species ever change into a new species, since they are thought to be reproductively isolated?

Well, there are actually plenty of *indirect* indications that speciation did in fact occur in nature. The diversity of species in the Galapagos Islands — sometimes called "Darwin's lab" — has become a classical example, especially the huge variety of finches spread out over the many islands. Interestingly enough, Darwin initially considered them just varieties and did not even mention on which island each variety occurred. So perhaps they are not really different species. How would we know? Taxonomy has its so-called "splitters" and "lumpers," who distinguish themselves either by splitting populations and species or rather by lumping them together. It is evident that Darwin had to be convinced by the "splitter" John Gould who told him that the specimens of mockingbirds Darwin had collected on three islands in the Galapagos were not just varieties but three distinct species. So, Darwin was not able to claim that they were a product of speciation, until Gould taught him that they were real species, not just varieties. That may seem self-serving.

Yet, from then on, Darwin would make the case that populations on separate islands may have been isolated long enough to develop major differences through natural selection, transforming them into different species, before related populations moved in; but once settled, they could no longer interbreed with the original population (alas, we were not there when it happened). Once a population has become reproductively isolated — according to the biological species concept — speciation has become a fact.

THE PROCESS OF SPECIATION

So the question remains: Is speciation possible? To begin with, it is very unlikely that a new species occurs at a single stroke, in one generation. There was not, for instance, a "first chimpanzee" whose parents were not chimpanzees. We are probably dealing with a gradual process. Biologists have been busy finding examples in nature of how this process works. Here is what they have come up with.

Speciation often starts with some form of geographical isolation. Once two populations of the same species become geographically isolated, mutations accumulate independently on both sides of the geographical barrier. In time, due to accumulating mutations, geographical barriers may lead to biological barriers. Take for instance the two fruit fly species *Drosophila pseudoobscura* and *Drosophila persimilis*; they are very closely related, yet isolated from each other by habitat (*persimilis* generally lives in colder regions at higher altitudes), by the timing of their mating season (*persimilis* is more active in the morning and *pseudoobscura* at night), by their behavior during mating (the females of both species prefer the males of their respective species), and by sterility of hybrid males. Even if the original physical barrier no longer exists, the two new species may stay reproductively isolated due to the biological barriers they have acquired. Once geographical isolation has created reproductive isolation, speciation has become a fact.

But there is more. Not all geographical isolation is based on physical barriers such as mountains, deserts, or water. There are also "barriers" of a different nature — such as the mere physical distance between some members of a population. In such cases, the population is still continuous, but nonetheless its members may not mate randomly because they are more likely to mate with their geographic neighbors than with individuals farther away from them. How could such a situation lead to speciation?

Let us consider a specific case where the species forms a ring of several neighboring populations who can interbreed with adjacent populations, but where there may be at least two "end" populations in the series that are too distantly related to interbreed. An example of this situation can be found in California, where the *Ensatina* salamander forms a horseshoe shape of populations in the mountains surrounding the Californian Central Valley.[4] Although interbreeding can happen between neighboring populations around the horseshoe, the salamanders on the

---

4   T. Dobzhansky in *A Century of Darwin*, ed. S. A. Barnett (Cambridge, MA: Harvard Univ. Press, 1958), 19–55.

western end of the horseshoe cannot—or perhaps we should say "can no longer"—interbreed with the salamanders on the eastern end. The problem, then, is whether to classify the whole ring as a single species (in spite of the fact that not all individuals can interbreed) or to consider each population as a separate species (regardless of the fact that there is still interbreeding with nearby populations). This much is clear: If enough of the connecting populations within the ring perish to sever the breeding in-between connections, the remaining members would have become two distinct species.

Doesn't this sound more like a semantic issue? Not really. It shouldn't surprise us that, in an evolutionary context, a clear-cut species concept inevitably becomes a bit fluid. The strength of the biological species concept is that it offers an explanation of what maintains and disrupts the unity of the species. To be more specific, it makes gene flow the binding force behind a species, and at the same time limits gene flow through isolating mechanisms. Once a species is in fact reproductively isolated, it receives a species name consisting of two parts—the genus name (starting with an uppercase letter) and the species name (starting with lowercase). For instance, *Panthera leo* is the species of lions, and *Panthera tigris* is the species of tigers; they both belong to the same genus, *Panthera*. Hybrids between them do exist, though, especially in zoos, but they are rare and often sterile.

Critics of Neo-Darwinism keep insisting that this view of speciation is not based on direct observation but merely on dubious assumptions and interpretations—all supposedly in line with the Neo-Darwinian axiom of continuous evolution, step-by-step, gene-by-gene, mutation-by-mutation. However there is also experimental evidence that speciation can and does occur this way. In the late 1980s, biologists bred fruit flies of the species *Drosophila melanogaster*, using a maze with different choices of habitat such as light/dark and wet/dry.[5] Each generation was placed into the maze, and the groups of flies that came out of two of the most extreme eight exits were kept apart to breed with each other in their respective groups. After thirty-five generations the two groups and their offspring had become reproductively isolated because of their strong

5 William R. Rice and George W. Salt, "Speciation via disruptive selection on habitat preference: Experimental evidence," *The American Naturalist*, 1988, 131 (6): 911–17.

habitat preferences: They mated only within the areas they preferred, and so did not mate with flies that preferred the other areas. The habitat split had actually become a reproductive split.

What seems to remain a problem, though, for the biological species concept is the existence of *hybrids*—crossings between the members of two different species. The best known examples are hybrids between a horse (with 32 pairs of chromosomes) and a donkey (with 31 pairs). They are called either a mule (if the father is a donkey or "jack") or a hinny (if the mother is a donkey or "jenny"). Female hybrids (both mule and hinny) can sometimes still produce offspring, but all male hybrids are infertile. What are we to make of such crossings in terms of speciation?

Ironically, hybrids are more of an anomaly for the static species concept than for the dynamic one based on the biological species concept. No wonder then, as we saw earlier, Thomas Aquinas and Carl Linnaeus did not quite know what to do with hybrids. Hybrids make sense only when seen from an evolutionary viewpoint, but they wouldn't make sense if all species were created once and forever. In terms of a biological species concept, hybridization is a phenomenon of speciation that has not yet come to completion. Hybridization may still occur after two populations of the same species have become separated and then come back into contact with each other. If their reproductive isolation had had a chance to become complete they would have already developed into two separate, incompatible species. But if, on the other hand, their reproductive isolation is not yet complete, further mating between the populations will still produce hybrids, although these may not be fertile. Obviously, species are very clear-cut and delineated in a static world, but in a world with evolution the species concept becomes a bit more fluid because there is speciation in progress.

The bottom-line of all of this is that Neo-Darwinism has a deep-seated belief in gradualism, which it inherited from Darwin. Darwin persisted in this belief throughout his life and passed it on to his successors:

> As natural selection acts solely by accumulating slight, successive, favourable variations, it can produce no great or sudden modification; it can act only by very short and slow steps. Hence the canon of 'Natura non facit saltum,' which every fresh addition to our knowledge tends to make more strictly

correct, is on this theory simply intelligible. We can plainly see why nature is prodigal in variety, though niggard in innovation. But why this should be a law of nature if each species has been independently created, no man can explain.[6]

6   Charles Darwin, *Origin of Species*, 1859, 471.

4

Evolution with Leaps

DOES NATURE REALLY MAKE NO LEAPS, AS DARWIN
suggested? Regarding this there have always been dissenting voices. In
the early days of evolutionary biology, when genetics was still taking its
first steps, some biologists began to speak of what they called "saltation."
Saltation (from Latin, *saltus*, "leap") is considered a sudden change from
one generation to the next—a large, or very large, genetic change compared to the non-gradual changes (especially single-step modifications)
typical of Darwinian gradualism. Ideas of saltation had been in the air
even before Darwin rejected them (e.g., Lamarck). But Darwin's complete
rejection of saltation would not satisfy biologists for long.

SALTATIONISM

Advocates of saltationism deny the Darwinian idea of a slowly and
gradually growing divergence of features as the *only* source of evolutionary progress. They did not necessarily completely deny gradual variation,
but would claim that drastically new "body plans" come into being as a
result of saltations — sudden, discontinuous, and crucial changes. The latter are supposedly responsible for the sudden appearance of new, higher
taxa, including classes and orders, while minor variation is supposed to
be responsible for the fine adaptations below the species level.[1]

In the early 20th century a mechanism of saltation was proposed based
on "super" mutations — large-scale mutations or quantum mutations. It
was seen as a much faster alternative to the Darwinian concept of a
gradual process of small random variations being acted on by natural
selection. This idea was popular with early geneticists such as Hugo de
Vries, William Bateson, and, early in his career, Thomas Hunt Morgan.
Some spoke of macro-evolution in contrast to micro-evolution. But at
the time it was hard to demonstrate such large-scale mutations.

Saltationism never became popular in evolutionary biology, mainly

---

1   G.S. Levit, et al., "Alternative Evolutionary Theories: A Historical Survey," *Journal
of Bioeconomics*, 2008, 10.1: 71–96.

because there was not much genetic evidence to base it on. In Darwin's time there was no field of genetics yet to even back up his claim of small modifications; nowadays genetics does support Darwin's claim but is only beginning to open the black box of sweeping genetic changes. Since then, genetics has been making progress, at an almost "saltational" speed. In 1984 the geneticist Barbara McClintock introduced the idea of "jumping genes," chromosome transpositions that can produce rapid changes in the genome, for which she would later receive a Nobel Prize.[2]

Perhaps the most important development in this connection was the insight that the world of chromosomes, genes, and DNA was much more complex than many geneticists had originally thought. The idea that genes are merely "beads" on strings of chromosomes turned out to be extremely simplistic. It has become common knowledge nowadays in genetics that genetic programs not only consist of "blueprint" genes — which contain DNA code for the production of proteins (enzymes are proteins too) — but also of instructions for processing the "blueprint." The Y-chromosome of males, for instance, hardly carries any "blueprint" genes for specific characteristics, yet this chromosome has quite an impact on the process of sexual differentiation, as we know from rare cases when it's missing. Clearly, we are still in largely uncharted territory here — with even the magnitude of the unknown being unknown as well! But geneticists are making progress, almost on a daily basis.

Perhaps the analogy of computer programming may help explain this new insight. If you were to print out computer code or script on a piece of paper, you would only see a bunch of letters and lines. But there is much more to it: structure. A line that performs a calculation, for instance, is very different from a line that contains a branch statement (e.g., "if X then Y") or a loop statement (e.g., "do X until Y"). Something similar is the case with DNA code. Genes come in at least two main forms, structural ones and regulatory ones (although some perform both functions). The structural (also called expressive or protein-coding) genes are those that create the cell proteins needed for structure and metabolism. But before a gene can be expressed, it may have to be "switched-on," and this is one of the functions of regulatory genes.

2   B. McClintock, "The significance of responses of the genome to challenge," *Science*, 1984, vol. 226: 792–801.

REGULATORY GENES

We mentioned already the existence of structural proteins and catalytic proteins, but it turned out there are also *regulatory* proteins. The genes for regulatory proteins produce either *activator* proteins or *repressor* proteins. These proteins can bind to certain sections within regular genes and thus activate or repress their activity.

That's how the protein production of a regular gene can be controlled — through the use of other proteins binding to short DNA-stretches within the gene itself. It's quite a complicated network of controls. We won't go into more detail here, for the entire picture is actually even more complicated, and keeps changing by the day.

Perhaps a few more developments in genetics should be briefly mentioned here. One of them is DNA methylation — a biochemical process whereby a methyl group is added to the cytosine or adenine DNA nucleotides. This may affect the activity of structural genes and thus alter the expression of those genes. Similar to this are histone modifications. Histones can package and order the DNA into structural units so they fit inside the cell nucleus, thus making the compacted molecule 40,000 times shorter than an unpacked molecule. They act as spools around which DNA winds, and as such can also play a role in gene regulation.

Another development relevant in this context is the discovery of the so-called *homeobox*, or *Hox*, genes. The morphological diversity of vertebrates, from humans to eagles, or from whales to snakes, evolved around a common set of developmental genes. Mammals, birds, and amphibians, for instance, share the same set of 39 *Hox* genes. It is these *Hox* genes that control the body plan of the embryo along the head-to-tail axis. In vertebrates, the various *Hox* genes are situated very close to one another on a chromosome in groups or clusters. Interestingly, the order of the genes on the chromosome is the same as the expression of the genes in the developing embryo. These genes somehow regulate that all vertebrates have a similar build; thus they make for the unity underlying morphological diversity. They are responsible for the large-scale segmental organization of vertebrate body plans.

These regulatory genes underwent a number of gene duplications, leading up to the separation of the arthropod and vertebrate lineages. In various arthropods the different appendices with leg-like shape could evolve into specialized antennae, claws, and many other structures sharing the same jointed architecture, but modified to serve quite different functions. Gene duplications of regulatory genes can apparently achieve this kind of evolution.

All these developments taken together explain why the human genome may not differ much from the genome of other species. A decade ago, the general estimate for the number of human genes was considered to be well over 100,000, since humans produce up to some 100,000 proteins. But the relationship between genes and proteins turned out to be far more complicated, because some genes can be overlapping and some can produce several proteins. So gradually the number of protein-coding genes has been lowered to around 21,000 — which is only a bit more than the 20,470 genes a tiny roundworm needs to manufacture its utter simplicity.[3] The idea that more complex organisms require more genes is, then, no longer valid. The water flea, for example, has 31,000 genes. And human beings have only 300 genes not found in mice. No wonder Craig Venter, former president of Celera Genomics, said about the determination of this surprisingly low number of genes for humans, "This tells me genes cannot possibly explain all of what makes us what we are."[4]

These new developments in genetics also give us reason to revisit the issue of saltations and quantum mutations. They may no longer be just phantom notions. In spite of what Darwin endlessly denies, nature does make leaps, and we can even see now why and how. Some leaps are so stubborn that they just cannot be ignored. Think of the leaps, or discontinuities, during the metamorphosis of a butterfly — from egg to caterpillar, to pupa, and finally to a mature adult. This kind of metamorphosis can be found in 80% of our fauna, and involves dramatic conversions with regard to physiological changes combined with structural body changes. Metamorphosis can even be regulated by one single gene. It has been shown for instance that transcription factors play key roles in juvenile hormone (JH) action and lead to the "leaps" seen in the changes towards a butterfly.

Apparently even small gene mutations can have large effects. Here is a classic example: There are two kinds of stickleback fish, one with and one without spike spines on the pelvis. About 10,000 years ago a

---

3 The task of spotting protein-coding genes is by no means easy. Iakes Ezkurdia, et al., have found indications that the number is even closer to 19,000 (https://arxiv.org/abs/1312.7111).

4 Quoted by Tom Abate, *San Francisco Chronicle* staff writer, Sunday, February 11, 2001.

mutation in a genetic switch near a gene involved in spine production differentiated the two varieties, into one with spines and one without, one adapted to oceans and the other to lakes.[5]

Let's take another example, the FOXP2 gene on human chromosome #7 — a member of the FOX group of transcription factors involved in regulation of gene expression. This gene was discovered in a family in England where members had great difficulty in speaking.[6] They had a severe speech impediment manifested in incomprehensible talk. As it turned out, these affected members had a single "letter" (nucleotide) of the DNA code "misspelled" in the FOXP2 gene.[7] Besides, two amino acid substitutions distinguish the human FOXP2 protein from that found in chimpanzees, but only one of these two changes is unique to humans. Not only do the human and chimp versions of FOXP2 look different, but they function differently as well. The FOXP2 gene switches other genes on and off, thus driving target genes to behave differently in both species. It could very well be that all these seemingly minor genetic changes had a major impact on the development of language, although there may be many other genes, and other factors, involved in language development (see chapter 6).

Sometimes genetic changes can be even more drastic. Here is a case of chromosomal changes. Chromosomes form when a cell divides into two new cells, which makes them visible through the light microscope. They are packed with DNA and may contain hundreds of genes. Each species has a certain number of paired chromosomes — one half coming from the female parent, the other half from the male. Chimpanzees, for instance, have 24 pairs of chromosomes, whereas humans have only 23 pairs. This change from 24 to 23 pairs of chromosomes can be explained if we assume that two ancestral chromosomes have fused together to form the human chromosome #2. This is even more likely to be the case, given that gorillas and orangutans also have 24 pairs, just as chimpanzees do, but unlike us. There is even more proof that chromosome #2 is a head-to-head fusion of two shorter chromosomes. The tips of

5   H. Allen Orr, "The genetic theory of adaptation," *Nature Reviews, Genetics,* 2005, 6: 119–27.

6   R. Nudel and D. F. Newbury, "FOXP2," *Wiley Interdisciplinary Review of Cognitive Science,* 2013, 4 (5): 547–60.

7   Francis Collins, *The Language of God* (New York: Free Press, 2006), 139–40.

all primate chromosomes have a special sequence of DNA code which is rare elsewhere in the chromosome. Well, this very sequence is found right in the middle of our fused chromosome #2. The fusion has left its DNA imprint behind, so to speak.[8]

If macro-mutations are possible at the species level, or between closely related species—and there are indications that this is the case—then we may assume they may also play a key role in more drastic transitions. What comes to mind is the transition from fish to reptiles (which is definitely more than changing swim bladders into lungs, or fins into legs), from reptiles to birds (which is, again, more than changing legs into wings), and so on. Here the gradualism of Neo-Darwinism seems to fail us. The Darwinian view claims that speciation is a universal, continuous, and gradual change ruled mainly by natural selection—thus making organisms look like "adaptation machines."

But the question is whether all forms in the natural world are fully determined by adaptation and natural selection. Evolution may have produced virtually endless variation, but it does so based on a rather limited set of basic designs. However those various structural designs share an even more basic design: the same genes for DNA-transcription, replication, nutrient uptake, core metabolism, and so forth—most likely inherited from a unicellular organism some 500 million years ago. Yet they are distinct in using distinct combinations of developmental programs.

8 Ibid., 137–38.

5

Humanity Step by Step

HOW DOES ALL WE DISCUSSED SO FAR APPLY TO THE evolution of *humanity*? Undoubtedly, humans have many similarities with the rest of the animal world, especially with the mammals: we have a backbone like they do, four extremities with the same basic pattern, seven cervical vertebrae, and the list goes on and on.

COMMON ANCESTRY

In addition, we also inherited features from the animal world that may be much harder to explain. The eye, for instance, is actually structured "inside out," with the nerves and blood vessels lying on the surface of the retina (instead of behind it, as is the case with many invertebrate species). As a consequence, this arrangement creates a blind spot and calls for a number of complicated adjustments, because light has to pass first through nerves and blood vessels in order to reach the light sensors.

And then again there is the curvature of the spine. Although humans stand on two legs, their skeleton was originally adapted to walking on four legs. Evolution accomplished this transition by breaking up the bow-curved arch of the back of four-legged animals into the s-curve we as bipeds now possess. The result is a set of creative adaptations, but accompanied by compromises. Back problems are the "scars" of this evolutionary past: strained muscles, slipped discs, and pinched nerves. Another outcome is that the long, flat pelvis of quadrupeds has become more basin-shaped, with a short, broad top bone — good for balance but bad for birthing because it narrows the birth canal.

There are also other "traces" left in our bodies that "remind" us of our ancestors in the animal world. The appendix, for instance, is the remnant of an organ once very functional when our ancestors were still dedicated plant-eaters. And our jaws, which are no longer as protruding as they used to be in our ancestry, have become overly crowded, often causing the third molar to cause trouble (our "wisdom teeth"). Or take the case of the human male, whose testes develop initially within the abdomen (another ancestral feature dating back to "cold-blooded" animals), but

33

which later on, just before birth, migrate through the abdominal wall into the scrotum, which is more beneficial for "warm-blooded" animals. No wonder, though, this process causes two weak points in the abdominal wall where hernias can later easily develop.

Even at the level of DNA there are many "traces" left of our ancestry. For instance we still carry pseudo-genes — structural genes that were once functional copies of genes but have since lost their protein-coding ability due to mutation. A striking example is the gene for a jaw muscle protein, MYH16, which has become a pseudo-gene in humans, but retains its function of developing strong jaw muscles in other primates. Another example would be the DNA sequence for an enzyme that produces ascorbic acid (vitamin C) in most animals. Many primates, including humans, have a defect in this DNA code, so they must acquire vitamin C through food, but they did hold on to its repetitive sequences in the "silent" section of their DNA.[1] Other examples of pseudo-genes in humans can be found in genes encoding distinct olfactory receptor molecules. A typical mammal has an estimated 1,300 of these genes, while this number was radically reduced in human evolution, where roughly 60% have degraded to pseudo-genes.[2]

These and many other similarities between humans and other animals are quite striking. True, having similarities does not necessarily mean having common descent, but it is the most probable explanation: Not only are we like them; we most likely also came from them. At this point, not quite surprisingly, Darwin's unshakeable belief in gradualism kicks in again. It is, in essence, an ideology of equality, assuming there are no "real" differences in the living world — no high and low, no discontinuities but just ongoing and gradual continuity, achieved with small incremental steps. Darwinism is essentially egalitarian in outlook and preaches equality, declaring that all organisms evolved equally as a result of small changes — in accordance with Darwin's famous statement: "the difference in mind between man and higher animals, great as it is, certainly is one of degree and not of kind."[3] In line with this doctrine,

1   I.B. Chatterjee, "Evolution and the biosynthesis of ascorbic acid," *Science*, 1973, 182: 1271–72.

2   T.W. Deacon, "Human Variability and the Origins and Evolution of Language," in *On Human Nature*, eds. Michel Tibayrenc and Francisco Ayala (San Diego: Elsevier, 2017), chapter 32.

3   Charles Darwin, *The Descent of Man* (Amherst, NY: Prometheus Books, Great Mind Series, 1998).

humans ought to be degraded to glorified animals, or reversed, animals ought to be inflated to pre-humans, humans-in-the-making so to speak. Thinking differently would be considered elitist in terms of the Darwinian ideology of gradualism.

GRADUALISM

When it comes to human genetics, anatomy, and physiology, Darwin's approach might be quite attractive. Guided by the Master's voice of gradualism, biologists have been searching for pre-human features in the animal world, especially in the world of higher primates. We will discuss their findings in upcoming chapters. But in this chapter we will specifically focus on what paleoanthropologists have been doing when studying fossils of our closest human and pre-human ancestors.

They came up with certain anatomical characteristics that are supposed to uniquely define human beings, and yet developed gradually from pre-human ancestors. One of the earliest criteria considered was brain volume. There is indeed a general increase in cranial capacity and brain size throughout the *Homo* lineage, from *Homo habilis* with 727–846 cc (about 2–2.8 million years ago) to *Homo erectus* with 850–1100 cc (about 1.9 million to 70,000 years ago), and expanding further from there. The explanation may be found in mutated enhancers located close to genes involved in building our brains. One of these enhancers, HARE5 (short for human-accelerated regulatory enhancer), differs between us and chimpanzees.[4] It appears that the human brain reached its current volume and size some 80,000 years ago.

When the brain volume of a skull was found to be above a certain minimum — some said 1,100 cc — paleoanthropologists were ready to assign the label *Homo sapiens* ("sapiens" is Latin for "wise"). This is a rather arbitrary criterion, though. *Homo sapiens* is indeed a "wise man" who surely has a relatively large brain compared to other mammals. However, brain volume can vary widely among human beings. Besides, brain mass or quantity is not clearly correlated to brain quality, let alone mental capacities. Neanderthals score relatively high on a brain volume scale — but so do elephants. What matters is not so much a bigger brain,

---

4   Lindblad-Toh, et al., "A high-resolution map of human evolutionary constraint using 29 mammals," *Nature*, 2011, 478: 476–82.

but whether it is bigger in the right spots and with the right connections. More generally, why would brain volume be considered such a defining property of human beings? It's probably because the brain is the organ where the mind is thought to be located; and the difference in mind between humans and other animals is, according to Darwin, one of degree and not of kind again. That's how brain size received its importance in the first place.

As a second move, some paleoanthropologists declared that each time a fossil showed signs of a creature capable of walking upright on two legs (biped), they take this as another kind of human-like qualification. Fossils showing evidence of bipedalism were assigned to the species *Homo erectus* ("erectus" is Latin for "upright"). Sure, being a two-footer seems to be a big plus for other human features — the main advantage supposedly being that the neck muscles put less restraint on the expansion of the brain part. But the question remains whether being a biped is really a decisive factor for being more human. It may not even be a necessary condition for being human, let alone a sufficient condition. Many animals are bipeds; all birds, all kangaroos, most dinosaurs, and even chimpanzees have been able to walk or hop pretty well on two feet. Does that really bring them closer to being human? Again, it's mainly the doctrine of gradualism that favors this feature.

Then there was a third move in paleoanthropology. Any kind of evidence of tool-using and tool-making was taken as another important indicator of human qualifications. The emergence of the first Oldowan stone technology is usually dated to two and a half million years ago. And so, paleoanthropologists assigned certain fossils to a new species, the species *Homo habilis* — the "handy man" ("habilis" is Latin for "handy"). Supposedly, the ability to use tools is facilitated by a so-called opposable thumb, which is a pad-to-pad kind of precision grip. It is not unique to humans, though, and can be found to different degrees among other Primates such as baboons; even opossums, koalas, and pandas have some form of "opposable thumb." An opposable thumb certainly makes it easier to *use* tools, and even more so to *make* tools.

Because making tools also requires a certain brain capacity, it seems fair to speak of a pre-human being when we find fossils accompanied by hand-made tools. However, pre-human ancestors started making stone tools as early as 2.6 million years ago. Besides, there is mounting evidence that chimpanzees and some other animals not only are able to use tools but also have some capability of making tools. On the other hand, tool-making could very well be a necessary but not sufficient condition for being a *human* tool-maker. The human mind of a tool-maker is able to create tools according to some mental concept of a tool object with which a continual comparison is made with the actual product being worked on. Thus they anticipate the future by means of abstract rational conceptions residing in their minds. We will discuss this issue more extensively in chapter 7.

What these three criteria — brain size, bipedalism, and tool-making — have in common is the assumption of gradualism: a gradual path leading from pre-human to human. Indeed, the brain does show a gradual enlargement; the spine shows a gradual change from a four-footer (J-curved) to a two-footer (S-curved); and the use of tools gradually develops into making tools. But because the transition is supposed to be gradual, it is by definition impossible to locate any clear-cut boundary lines in this process. The doctrine of gradualism in essence denies that there is any clear transition point from pre-human to human.

An added problem from human paleontology is that most finds and excavations represent only a single individual. Given the fact that specific morphological measurements may vary widely among the members of a population or species, a single find may not be very representative for the population or species it belongs to. This could easily make for anecdotal or circumstantial evidence, which certainly casts doubt on extravagant claims about the entire population or species. Very detailed and precise measurements of skulls and bones may give us the impression we are dealing here with an exact science, but what paleoanthropologists derive from these exact measurements is not very "exact" at all. G. K. Chesterton once skewered the pretensions of anthropologists who spun detailed theories about the culture and capabilities of primitive man: "We have a series of hypotheses so hasty that they may well be called fancies.... The professor with his bone becomes almost as dangerous as a dog with his bone."[5]

This brings us to another problem in the context of paleontology. How do paleontologists assign a fossil to a particular species? What is it that makes certain fossils belong to the species *Homo erectus* and others to

[5] G. K. Chesterton, "The Everlasting Man," in *Collected Works of G. K. Chesterton* (San Francisco: Ignatius Press, 1987), book 2, part I, ch. 2 ("Professors and Prehistoric Men").

the species *Homo sapiens*, for instance? The species concept is in any case rather controversial in evolutionary biology, as we discussed earlier (see chapter 3). For a long time a species was considered to be a collective name for a group of animals sharing the same characteristics and thus meeting the requirements to be assigned to the same category, in the way Linnaeus did. This approach led to what is called a "typological" species concept. Even if we want to use this concept in paleontology, the problem is that it is difficult to determine what is "typical" for a population or species, given the fact that paleontology has only a few fossils available. It is hard to treat a specific fossil as representative of the population or species when not much is known about other, missing representatives and their variability. Rather telling is the fact that some experts considered one of the first Neanderthal remains — discovered in 1856 at Germany's Feldhofer cave — to represent a diseased anomaly; the famous German scientist Rudolf Virchow, for example, claimed that the Neanderthal specimen he examined had rickets and arthritis.[6] Besides, when comparing a sparse collection of skulls with each other it is rather arbitrary to determine whether they are "similar" enough to be assigned to the same species.

On the other hand, although the biological species concept may be a better one, it cannot work well for paleontologists either, as paleontology is hardly an experimental science, for there is no way to check for reproductive borders. Yet paleontologists do assign specific fossils to specific species — which potentially makes their claims somewhat problematic. Comparing skeletons, skulls, and teeth — and then classifying them — is inevitably a rather intuitive approach with wide margins of error. True, currently there is more DNA information available, but this doesn't work miracles either. DNA analysis does not give us a clear portrait of the carriers of this DNA. We still have very little knowledge of what most DNA segments do and of how they make a human being into the person he or she becomes. Chesterton sets things right again: "Strictly speaking of course we know nothing about prehistoric man, for the simple reason that he was prehistoric. The history of the prehistoric man is a very obvious contradiction in terms."[7]

6 Adam Kupler, *The Chosen Primate* (Cambridge, MA: Harvard University Press, 1996), 38.
7 Ibid.

Nevertheless, scientists have no other way than working with physical or material characteristics. In that specific sense, Darwin was arguably right when he said that the difference between animals and humans is one of degree, not of kind, especially so when it comes to their anatomy. Indeed, since we are "made of their cloth," during evolution there must have been some features in the animal world that came closer and closer to features more characteristic of the human world — features such as making tools, walking upright, and having a relatively large brain volume. In a purely biological sense there is a gradual transition; so, seen from a biological point of view, we surely are "glorified animals," as animals are "humans-in-the-making." A gradual evolutionary process was somehow able to prepare the way for the appearance of humanity.

Regardless of how this gradual transition in human evolution took effect, the next two pivotal questions would be: Did humanity arise at one time in one place (monophyletism) or rather at different times in different places (polyphyletism)? And secondly: Did humanity originate from one or two individuals (monogenism) or rather from a large pool of individuals (polygenism)?

At one point several paleoanthropologists favored polyphyletism, but that opinion has few defenders left. Thanks to scientific advance, the dominant view now is that all human beings in fact descended from common ancestors living some 150,000 years ago in East Africa.[8] The "out-of-Africa" model, which proposes a single area of origin for modern humans,[9] is the current dominant model of the geographic origin and early migration of anatomically modern humans (*Homo sapiens*). Interestingly, this idea is much closer to what the Catholic Church has always taught. The "unity of the human race" might be called an "indirect" dogma in Catholicism, since it is necessarily presupposed by the doctrine of Original Sin, which Chesterton once called "the only part of Christian theology which can really be proved."[10] It is hard to see how every human being who ever lived could have shared a common fallen

---

8 This date has recently been pushed farther back to between 177,000 and 194,000 years ago with a finding in an Israeli cave near Carmel. Israel Hershkovitz, et al., "The earliest modern humans outside Africa," *Science* vol. 359, issue 6374 (Jan. 26, 2018): 456–59.

9 H. Liu, et al., "A geographically explicit genetic model of worldwide human-settlement history," *Am. J. Hum. Genet.*, 2006, 79 (2): 230–37.

10 G. K. Chesterton, *Orthodoxy* (Chicago: Moody Publishers, 2009), chapter 2.

human nature if polyphyletism held true. For example, if the historic event of the Fall had taken place in Africa, how could this event have affected the human natures of the individuals who evolved independently somewhere else?

The discussion as to the second question — did humanity originate from one or two individuals (monogenism) or from a gene pool of individuals (polygenism) — is far less settled. In evolutionary biology the unit of evolution is the population, not the individual — think of the gene-pool model (see chapter 2). Whereas organisms are the units of selection, they are not the units of evolution — populations are. Organisms do not evolve but populations do. So the idea that humanity originated from one or two individuals is seemingly hard to reconcile with biology (see chapter 3). Population-thinking simply precludes the possibility of a founding organism for a biological species.

The legendary geneticist and evolutionary biologist Theodosius Dobzhansky had already expressed and "endorsed" the idea of polygenism:

> Since species differ in numerous genes, a new species cannot arise by mutation in a single individual, born on a certain date in a certain place. [...] Species arise gradually by the accumulation of gene differences, ultimately by the summation of many mutational steps which may have taken place in different countries and at different times.[11]

Even the Jesuit paleontologist Teilhard de Chardin was of this opinion: "Thus in the eyes of science, which at long range can only see things in bulk, the 'first man' is and can only be a crowd, and his infancy is made up of thousands and thousands of years."[12] Apparently, Teilhard thought he had to go for polygenism (he even slightly tended towards polyphyletism). He thought that a scientist cannot directly address the hypothesis of monogenism — and certainly not the idea that humanity had started from one individual, Adam. He argued that all we believe we know about biology renders the hypothesis of one or two individuals

11 T. Dobzhansky, *Mankind Evolving* (New Haven, CT: Yale University Press, 1962), 180–81.

12 Pierre Teilhard de Chardin, S.J., *The Phenomenon of Man* (New York: Harper and Row, 1959), 105.

untenable. In his own words, "It is irreconcilable with what we know from biology that our human species should be descended from a pair."[13]

Based on this idea, we are given the impression by most paleoanthropologists that the ancestral population of humans never shrank below a number of approximately 10,000 individuals.[14] This explains why they portray our history as a biological species shaped by migration, interbreeding, and unrelenting adaptation, which has generated much diversity within the human population. We will discuss later whether this view can still be defended.

To sum up, biology has indeed given us much more information about the origin of humanity. Through new advances in DNA research biologists are now even able to trace back what the DNA of our (first human?) ancestors must have looked like. This is done by comparing the non-coding ("silent") DNA sections of two types of DNA: mitochondrial DNA and Y-chromosomal DNA. What kind of DNA is that? All mitochondria in the egg-cell come from the mother only, since sperm-cells have no mitochondria. Since mitochondrial DNA (mtDNA) does not get "reshuffled," the mitochondrial DNA in the offspring is identical to that from the mother. In other words, mitochondrial DNA is passed unmixed from a mother to all her children, along the maternal line. Something similar holds for Y-chromosomal DNA as well. Since there is no counterpart to the Y-chromosome in the egg-cell, it does not get "reshuffled" either, so the Y-chromosome in the son is an identical copy of the Y-chromosome in the father. It is passed unmixed from fathers to all their sons, along the paternal line.

Either kind of DNA can be tested for single-nucleotide-polymorphisms (SNPs), which are single base pair changes in the DNA. Any specific inherited SNP in non-coding DNA sections creates a "DNA signature" (*haplotype*). Since SNPs can last through thousands of generations, they can be used as markers to trace an individual's ancestry. A group of individuals who share similar haplotypes is known as a *haplo*-group, identified by certain DNA markers of rare mutations in non-coding DNA segments. All individuals in a given *haplo*-group have a common

13 As quoted by Malachi Martin, *The Jesuits* (New York: Simon & Schuster, 1988), 287.
14 E.g., H.L. Kim, et al., "Divergence, demography, and gene loss along the human lineage," *Philosophical Transactions of the Royal Society B: Biological Sciences*, 365 (2010): 2451–57.

ancestor at some point in time. Each *haplo*-group can then further split into subgroups characterized by some additional markers.

In 1987, after some calculations, it was concluded that "Y-chromosomal Adam" lived about 60,000–90,000 years ago in Africa.[15] The date is rather approximate, since such calculations are not very exact due to some uncertainty about mutation rates.[16] In a similar way, everyone alive today can also be linked back to "mitochondrial Eve." She must have lived about 140,000–120,000 years ago. But as usual in science, things keep updating. More recently, the difference in years has been estimated for "Adam" between 120,000 and 156,000 years ago, and for "Eve" between 99,000 and 148,000 years ago.[17]

Whatever the outcome is, "mitochondrial Eve" does not seem to be contemporary with "Y-chromosomal Adam." This may look odd, but there are several explanations. Perhaps the best one runs like this. If a disaster were to wipe out humanity except one man — let us call him Noah — plus his wife, his sons, and their wives, these people would be the ancestors of all descendants coming forth from them. Now this Noah would be the most recent common male ancestor through a strictly male lineage — he would be the "Y-chromosome Adam," so to speak. But "mitochondrial Eve" would not be Noah's wife, but the most recent common female ancestor of his daughters-in-law through a strictly female lineage. This female ancestor must go much further back than Noah himself.

It is through this kind of DNA analysis that we can construct a DNA family tree that has at its origin a "mitochondrial Eve" and a "Y-chromosomal Adam" representing our most recent common ancestors that have DNA markers like ours. Such a reconstruction tells us that these ancestors were living somewhere in Africa — which is beautifully consistent with the out-of-Africa theory and monophyletism. All other contemporaries of these "proto-humans" somehow failed to produce a direct unbroken male or female line to present-day human populations.

It is important to keep in mind that "mitochondrial Eve" and "Y-chromosomal Adam" are merely the most recent common matrilineal

---

15   R.L. Cann, et al., "Mitochondrial DNA and human evolution," *Nature* 325 (1987): 31–36.

16   Very roughly every 125 years.

17   G.D. Poznik, et al., "Sequencing Y Chromosomes Resolves Discrepancy in Time to Common Ancestor of Males Versus Females," *Science* 341 (2013): 562–65.

and patrilineal ancestors—they are more of a mathematical than a biological nature. They do not represent the first beginning of humanity,
but merely stand for a construct of our two most recent common representatives in a matrilineal or a patrilineal way. We should certainly
not confuse them with Adam and Eve as mentioned in the Bible—that's
an entirely different story. Genetics studies the *Book of Nature* in which
"Y-chromosomal Adam" and "mitochondrial Eve" each represents a
branching-point in the tree of evolution, whereas the Judeo-Christian
religion studies the *Book of Scripture*, in which Adam and Eve represent
the first parents of humanity, who started the first dysfunctional family,
which developed a broken relationship with God. We cannot identify or
equate them; we cannot read the *Book of Scripture* as if it were the *Book
of Nature*, nor the other way around. The *Book of Nature* tells us where
we come from in a biological sense, whereas the *Book of Scripture* tells
us where we come from in a religious sense.[18]

What biology did find out for us is that no species is more unlike
its ancestors than is the human species. Analysis of the human genome
shows us that chimpanzees are two to four times more genetically diverse
than the global human population—which tells us that humans of the
species *Homo sapiens* are all extremely closely related to each other, much
more so than even our closest living non-human relatives. This is certainly
a confirmation of the "unity of the human race." Interestingly enough,
the "unity of the human race" is what might be called, as we said earlier, an "indirect" dogma of the Catholic Church, since it is necessarily
presupposed by the doctrine of Original Sin.

But there is also another dimension to human uniqueness, for genetic,
physiological, and anatomical features are not the determining factors for
being human. Although we do breed, feed, bleed, and excrete as other
creatures do—yes, we are connected all the way down—what does in
fact set us apart from the animal world is something not necessarily of
a biological nature, namely our faculties of language, rationality, morality, self-expression, and religion—a topic that will concern us in the
next chapters.

But before we go there, let us stress one more time that all previous
discussions were based on the idea of gradualism. It is an "idea" that has

18 This distinction goes as far back as Augustine: *Enarrationes in Psalmos*, 45, 7.

permeated all the previous chapters. How should we assess this idea? On the one hand, it could be seen as a paradigm. The term "paradigm," which Thomas Kuhn introduced, has been defined in many ways, but typically it stands for a collection of rules on how to solve scientific puzzles.[19] Most scientists feel attached to the paradigm they were brought up with. The reason for this is that individual scientists acquire knowledge of a paradigm through their scientific education and training. They acquire their standards by solving "standard" problems, performing "standard" experiments, and eventually doing research under a supervisor already skilled within the standard, shared paradigm. Aspiring scientists become gradually acquainted with the methods, techniques, and presuppositions of that particular paradigm. They become part of a "school." For most biologists that school is no other than Neo-Darwinism.

On the other hand, gradualism could also be characterized as a paradigm that in fact requires quite a dose of "faith," rather than scientific evidence. However, it becomes a doctrine when any other ways of interpreting data are pre-emptively blocked. When that happens, gradualism declares with unwarranted "authority" that there are no revolutionary leaps in the evolution of humanity—and even that there can't be any. According to gradualism as a doctrine, all of evolution is supposedly accomplished gradually and slowly, with short steps and small mutations, one step at a time. But how could we ever know—*a priori*, that is—that gradualism is the only possible way of developing new features in evolution? In the upcoming chapters we will discuss this issue further by asking the question: Is gradualism really the only possible or legitimate way for evolution to occur?

19 Thomas S. Kuhn, *The Structure of Scientific Revolutions* (Chicago: University of Chicago Press, 1996, 3rd edition), 10.

6

The Language Divide

HUMANS ARE MASTERS OF ANTHROPOMORPHISM, SO
they tend to think that animals have got to be like humans, even when
the differences are quite obvious. Darwin's gradualism has only strength-
ened this idea. If we have language, animals must have language too;
if we have rationality, animals must have something like it; if we have
moral rights, animals must have them as well; and the list goes on and
on. To think differently is considered as adopting an unjustified feeling
of superiority. Not only that, but it goes against the doctrine of gradualism — which is "anathema" for those biologists who were indoctrinated
in the paradigm of gradualism.

ANTHROPOMORPHISM

The idea that if humans have language then non-human animals must
have language also is actually rather old. The legend of "speaking" animals
is widespread: non-human animals who can produce sounds or gestures
resembling those of a human language are to be found all over the globe.
There are many examples: the donkey of Balaam in the Book of Numbers
(22:28–30); Reynaert the Fox; the speaking animals of Aesop's Fables; a
talking lion by the name of Aslan in the world of Narnia; Donald Duck
and Mickey Mouse featuring in Walt Disney's Productions; and the list
goes on and on. The culmination was in the classic 1932 film *Tarzan the
Ape Man*, portrayed with a fanciful dialogue between Jane and Tarzan:
"Me Tarzan, you Jane."

We seem to be mesmerized by "speaking" animals. Yet deep down
we know they are not humans and cannot really speak. These animals
merely figure in legends, fables, and cartoons. We know that if there
is something animals cannot do, in real life, it's speaking, because they
don't have language! Although dogs can bark in different ways and
birds can sing with a rich repertoire of songs, none of them is able
to speak a language — perhaps they make sounds as we do, but that
is not the same as using language as we do. Even parrots may sometimes sound like us, but they are not using language as we do — they

merely imitate the sounds of human speech, without knowing what those sounds mean.

Nevertheless, Darwin's legacy of gradualism is still very powerful. He made us look at human beings as "glorified animals," who inherited all they have from the animal world—even language. One might wonder though whether our inability to let animals be animals has something to do with our inability to let human beings be human beings. Yet, we don't want to think in terms of discontinuity and superiority—for we are supposedly just modified animals. No wonder, then, that biologists have been searching for the animal in us—that is, for animal roots of our more specific human features. And language is one of the more important human features they like to go after. The search for speaking animals has been on ever since!

The more recent heroes in this pantheon of speaking animals are the world famous chimpanzees Sarah and Washoe, as well as the equally amazing "linguist" Koko the Talking Gorilla, and most recently the amazing Nim Chimpsky, the Talking Chimpanzee. The late David Premack, who was Professor of Psychology at the University of Pennsylvania, was the one who taught Sarah the chimp to communicate by using cards. And the Gardner duo, Allan and Beatrix, taught Washoe the chimp to use gestures inspired by the American Sign Language. Francine Patterson taught Koko the gorilla to understand more than 1,000 signs of what she calls "Gorilla Sign Language." And Herbert Terrace and his team made the chimp Nim learn 125 signs of American Sign Language.

Looking back now, the verdict on all these trials is rather detrimental to what they wanted to claim. Although the chimp Nim was able to learn 125 signs of American Sign Language, he never got beyond memorized two-word combinations lacking any syntactical structure.[1] The linguists Robert Berwick and Noam Chomsky say about Nim, "All that Nim was actually able to learn about American Sign Language was a kind of rote memorization—(short) linear sign sequences. He never progressed to the point of producing embedded, clearly hierarchically structured sentences, which every normal child by age three or four can do."[2] Even Premack

---

1 For a good critical analysis, see H. Terrace, *Nim: A Chimpanzee Who Learned Sign Language* (New York: Alfred Knopf, 1979).

2 Robert Berwick and Noam Chomsky, *Why Only Us: Language and Evolution* (Cambridge, MA: MIT Press, 2016), 145.

himself, the "grand old man" of behavioral studies on primates, eventually altered his originally positive views about Sarah's language abilities. After more than 25 years of research on the origin of language in the animal world, Premack was forced to come to the conclusion that the emergence of human language is "an embarrassment for evolutionary theory."[3]

LINGUISTICS

Attacks came also (predominantly) from experts in the field of linguistics. They pointed out early in the discussion that the simple "sentences" apes like Washoe, Koko, and Nim seem to form turn out to lack any kind of grammar—they are something like "me Tarzan, you Jane." But we all know that the difference between "Help me accept" and "Accept my help," or between "Your trouble" and "You're trouble," or between "Be well, so you can do good" and "Be good, so you can do well" is a matter of . . . well, grammar.

Language comes not only with semantics, but with a syntax as well—also known as grammar. Take the word "ape." Not only does it have a "meaning" and a "reference"—its semantics—but it is also a noun, which determines its specific position in a sentence—its syntax. It can be preceded by an article such as "the," as in "the ape." Between article and noun, an adjective can be inserted: "the smart ape." After the noun, a preposition can be added: "the smart ape of Premack." No matter what the noun stands for, it is open to any of the above combinations. Next, the combination "the smart ape of Premack" can become part of a larger combination: "the amazing activities of the smart ape of Premack." And this can go on indefinitely, which is called "recursive generation." So instead of focusing on words (semantics), we also need to focus on the relationships between words (grammar) when we talk about language.

Language is often seen as a collection of words that are considered "labels" for something else behind the words—that something being concepts (the "meaning" of the word) and objects (the "reference" of the words). Based on this idea we can indeed say, "What's in a word?" But language is more than a communication tool that employs labels connected to concepts and objects. Words in human language are not

3   David Premack, "'Gavagai!' or the Future History of the Animal Language Controversy," *Cognition*, 19 (1985): 281–82.

just "labels." True, animals may very well be able to use labels—as Sarah the chimp did with cards—but whether they can use actual language is a completely different issue. Apes may be able to link words or visual symbols to objects, but they have never been able to link them to other symbols in a grammatical, recursive, and structured way.

It's striking to note how humans almost "instinctively" use language, whereas apes do not. Apes have to be *forced* to "say something," whereas children develop this capacity naturally (otherwise we would call upon a therapist for help). As far as we know, the "language faculty" is uniform in the human species (apart from pathological exceptions). A newborn human infant instantly selects from the environment *language*-related data, whereas an ape, with the same auditory system, picks up only noise. While apes hear nothing but noise, infants extract language-relevant material from the noise. The newborns of English-speaking mothers are even able to distinguish the melodic structure of French from that of English, and vice-versa.[4] We may also expect newborns from Nigeria or Papua, New Guinea to easily learn the English language if we bring them to New York City. Besides, it is striking to see that the diversity of the world's 7,000 or so languages is restricted to a few structural variants.

Yet Darwin's gradualism keeps haunting biologists. In his own words, "natural selection can act only by taking advantage of slight successive variations; she can never take a leap, but must advance by the shortest and slowest steps."[5] So Neo-Darwinian biologists believe—or are supposed to believe—that human language must have its origin in the pre-human animal world. Since there are plenty of communication systems in the animal world, these biologists assume language *must* be for communication of the same sort as other communication systems in the pre-human animal world. Again it is gradualism that makes them believe that our language capacity *must* have evolved from animal communication.

It is this very assumption that has been challenged lately, particularly by the work of Berwick and Chomsky, both specialized in the field of linguistics. Their thesis is that language is not primarily a system of

---

4   J. Mehler, et al., "A precursor of language acquisition in young infants," *Cognition* 29 (1988): 143–78.

5   Charles Darwin, *Origin of Species*, 1859, 194.

communication but a system of *thought* — which can also, but only secondarily, be used for communication. They emphatically point out that all animal communication systems differ radically from human language. What, then, makes human language so different from (other) communication systems?

According to Berwick and Chomsky the fundamental difference can be found in the *hierarchical* structure of human language. They introduced what they call *Merge* as the single operation that builds the hierarchical structure required for human language syntax. *Merge* is to human language syntax what the "CPU" is to a computer. In their own words, "This operation takes any two syntactic elements and combines them into a new, larger hierarchically structured expression."[6] For example, two items such as *the* and *bananas* are assembled into the set {the, bananas}. Crucially, *Merge* can then be applied to the results of its own output, so that a further application of *Merge* to *ate* and {the, bananas} yields the set {ate, {the, bananas}}. In this way is built up the full range of hierarchical structure that distinguishes human language from all other known non-human communication systems.

A hierarchical structure may sometimes look very deceiving, as if it is merely a simple iteration of thoughts about thoughts: from "I think that P" to "I think that you think that P," to "I think that you think that Darwin thinks that P," and so on. This kind of iteration may seem *linear* but it is also *hierarchical*. Here is a more complex example: "How many books did you ask your children to ask their teachers to ask their bookstore to buy?" There can of course be many more levels in the hierarchy — in fact infinitely more. But there is something striking here: No matter how many levels there are in the hierarchy, each level determines what belongs together. In our example, the word "buy," for instance, is not connected with bookstores or teachers or children, but instead with "books," on the same level in the hierarchy — in spite of the fact that "books" and "buy" are the farthest apart from each other in linear distance. And the word "their" in front of "bookstore" does not refer to the children's bookstore but to the teacher's bookstore. It's all about structure!

Somehow, all language users are able to understand such sentences and to make the right connections. To show very convincingly that it is

---

6 Robert Berwick and Noam Chomsky, *Why Only Us*, 10.

not *linear* distance that matters in human language but only *structural* distance, Berwick and Chomsky use the following example: "Instinctively birds that fly swim."[7] How do we know in this sentence that "instinctively" modifies "swim," not "fly"—even though "instinctively" is closer to "fly" in terms of linear order? The answer can be found in the "depth" of the hierarchical structure: it is closer to "swim" in terms of structural distance—"swim" is embedded one level down from "instinctively," but "fly" is embedded two levels down from "instinctively." More formalized: {instinctively, {{birds, {that,fly}}, swim}}. Somehow, language users are able to make these hierarchical connections. Experiments have shown that children understand that syntax rules are structure-dependent as early as they can be tested—that is, by about age three.

This ability to create hierarchically structured language seems to be unique to human beings. To some degree, non-human animals are able to string items together serially and sequentially, but not in a hierarchically structured way. Animal communication systems are always based on "linear order." Birdsong, for instance, does not have a hierarchical structure as language does—birds don't sing motifs within motifs, for instance. Attempts to teach Bengalese finches a song with "hierarchical syntax," instead of "linear order," have failed. But even one of our closest "relatives," the chimpanzee Nim, could only memorize some two-"word" combinations, and thus never came close to the hierarchical structure of even the simplest sentence.[8] So it looks as though we may confidently conclude that no animals are able to pass the "language test"—only humans do. As far as language is concerned, there is a deep divide between them and us. Capacity for language seems to be quite a "leap" in the presumably gradual evolution from pre-humans to humans.

SPEECH

But there is obviously more to human language than syntax. In addition to *Merge*, language requires two other components, so we end up with three components in all: that is, a combinatorial operator for word-like atomic elements (*Merge*), and two interfaces: the conceptual system (for thought) and the sensorimotor system (for vocal learning

---

7 Ibid., 117.
8 Charles Yang, "Ontogeny and phylogeny of language," *Proceedings of the Natural Academy of Sciences of the USA*, 2013, 110 (16): 6324–27.

and production).[9] The former interface (for thought) we will discuss in chapter 7, but the latter one deserves attention right now.

It is the sensorimotor part that has deceivingly led some biologists to believe that language is primarily for communication and has gradually evolved from communication systems in the animal world—a very questionable assumption indeed. It is surely true that in speech we do impose *linear* order on words, because words can only be spoken in a temporal sequence. But language itself is not ordered linearly but hierarchically, as we have seen. It has a different function, one not primarily based on communication. It can be used for communication, of course, but so can gestures. And it can also be used for much else besides, of which thought is the foremost. Language merely uses the sensorimotor system as a tool to express thought. It is first a cognitive tool before its role as a communication tool comes into play.

The sensorimotor system has little to do with language; it actually evolved much earlier and is thus shared with other animals. In other words, the sensorimotor system was "language ready" before language emerged. It seems to have been basically intact for hundreds of thousands of years, whereas the language system itself is a rather recent development, unique to the species *Homo sapiens*. It was only at some later stage of evolution that the internal language of thought was connected to the sensorimotor system. That's when thought became connected to speech. But as Alvin Plantinga, philosopher at Notre Dame University, rightly remarked, speech is an activity of mouth and vocal cords rather than of the brain: "Your mouth speaking is dependent on appropriate brain activity; it hardly follows that speaking is just an activity of your brain."[10]

Not surprisingly, then, linear sequencing is also found in songbirds, and inevitably in human speech, but not in human language. Therefore, Berwick and Chomsky emphatically declare, "birdsong is only a model for speech, if that—not language."[11] This leads them to say that Aristotle's classic dictum about language—"Language is sound with a meaning"—should actually be reversed: "Language is meaning with sound."[12] In other words, "meaning" was established first in human

---

9 Robert Berwick and Noam Chomsky, *Why Only Us*, 40.
10 Alvin Plantinga, "Against Materialism," *Faith and Philosophy*, vol. 23 (Jan. 2006): 23.
11 Robert Berwick and Noam Chomsky, *Why Only Us*, 140.
12 Ibid., 101.

evolution—sound came later. In that sense, the first humans were "think-ers" before they were "talkers."

All of this indicates that language—instead of starting as a commu-nication system—evolved more like a "tool for thought."

> The most elementary property of our shared language capacity is that it enables us to construct and interpret a discrete infin-ity of hierarchically structured expressions: discrete because there are five-word sentences and six-word sentences, but no five-and-a-half-word sentences; infinite because there is no longest sentence. Language is therefore based on a recursive generative procedure that takes elementary word-like elements from some store, call it the lexicon, and applies repeatedly to yield structured expressions, without bound.[13]

WHEN AND HOW

This raises the question as to when and how language did emerge in evolution. To be more precise, it is not language itself, but the faculty of language, that emerged at some point in time. But *when* did that happen? If language is indeed primarily a "tool of thought," we may assume it was present when the first symbolic expressions were found in archeol-ogy. The earliest symbolic artifact found so far is a piece of ochre with a crosshatch design carved in it, dated to about 80,000 years ago, found in the Blombos Cave in South Africa.[14] It seems likely that language was involved there. If so, the faculty of language could have emerged some 80,000 years ago, and probably quite suddenly—although this is of course a controversial claim in the eyes of those in the gradualism camp.

The people of the Blombos Cave showed clear signs of symbolic behav-ior—and therefore, most likely, of language—before their exodus to Europe. Something must have set them apart from the Neanderthals, who did make artifacts, but most likely not symbolic art. Neanderthals were already around before humans had left Africa to spread around the world. This indicates that those humans leaving Africa already had language. The split between Neanderthals and us has been dated to roughly 500,000

---

13   Ibid., 66
14   Christopher Henshilwood, et al., "Emergence of modern human behavior: Middle Stone Age engravings from South Africa," *Science*, 2002, 295: 1278–80.

years ago. Quite soon thereafter, Neanderthals would migrate to Europe, so that they were most likely not even around when anatomically modern humans emerged in Africa. There is no real evidence that Neanderthals also had the rich symbolic life we find in the Blombos Cave. As a matter of fact, there are crucial differences of developmental nervous system genes between humans and Neanderthals.[15] These changes in regulatory genes most likely led to human brains winding up more globular in shape, over a longer childhood time span, than those of Neanderthals. As a result, Neanderthals have a "bulge" at the back of the brain, whereas in humans the increase in cranial capacity is shifted to the front. We know for instance that the gene MCPH1 regulates brain size and that one genetic variant of it in modern humans arose some 37,000 years ago.[16]

As to *how* language emerged, we are (still?) very much in the dark. This question would be easier to answer if language had developed from a pre-human communication system evolving into a human communication tool, but that is not likely to be the case if the previous considerations are true. To narrow our discussion to the "CPU" of language, we need to find out how the capacity of *Merge* could emerge. Are there any genes involved, and if so, which mutations made this happen?

Yes, there are some vague indications that there are genes for language use. In 1990, geneticists at the Institute of Child Health in London first reported a speech disorder that appeared in three generations of Britons known as the KE family.[17] They found 15 affected family members who seemed to have inherited problems with grammar, syntax, and vocabulary that were tied to poor control of facial muscles and difficulty pronouncing words. Although it seemed clear that there had to be a genetic link, researchers hunted for more than a decade before they could identify the gene believed to be responsible.

The gene was finally found in a fetus that had a translocation in chromosome 7, and who then later developed speech and language problems strikingly similar to those seen in the KE family. The translocation found

15 Mehmet Somel, et al., "Human brain evolution: Transcripts, metabolites and their regulators," *Nature Reviews Neuroscience*, 2013, 114: 112–27.

16 Patrick D. Evans, a.o., "Microcephalin, a Gene Regulating Brain Size, Continues to Evolve Adaptively in Humans," *Science*, Sept. 9, 2005, vol. 309, issue 5741, 1717–20.

17 Kate E., Watkins, a.o., "Functional and Structural Brain Abnormalities Associated with a Genetic Disorder of Speech and Language," *The American Journal of Human Genetics*, 1999, 1215–21.

in the boy had disrupted a gene called FOXP2—which we mentioned earlier (see chapter 4)—and they found the same mutation in the 15 members of the KE family. It must be noted however that FOXP2 does not single-handedly wire the brain for language. It is, rather, a transcription factor that turns other genes on or off by telling them whether to transcribe their DNA into messenger RNA, which then leads to the production of certain proteins. As a result, FOXP2 has a broad repertoire in embryonic development, playing critical roles also in the formation of the lungs, heart, and intestines.

So do we have a "language gene" here? Not really. FOXP2 is a gene essential for vocally expressing language (speech), but probably not for language itself. It is primarily a part of the sensorimotor interface—like the printer attached to a computer rather than the computer's CPU. As we found out already, this sensorimotor system is not human-specific. We find it already with vocal-learning song-bird species such as the zebra finch and the hummingbird. FOXP2 targets other genes such as ROBO1, which may cause dyslexia and speech disorders in humans when mutated. As said earlier (see chapter 4), it has been found that there are two small amino-acid differences between the protein that human FOXP2 codes for and that of other primates and non-human animals. So it seems fair to conclude, "FOXP2 cannot be regarded as 'the' gene 'for' language, since it is only one of many that have to be functioning properly to permit its normal expression."[18] It is at best a necessary but not sufficient factor.

So our original question remains unanswered. All we can add to the discussion is that specific areas in the brain—Broca's area and Wernicke's area in particular—have been associated with language, or more particularly with speech. But again, that's for the expression of language, not necessarily for language itself. Although it has been claimed that no "new" brain structures evolved to support language use, language functions may have effectively "recruited older neural systems, previously adapted to serve other functions...with old structures performing unprecedented new tricks."[19] Yet the language function is not some prior function that requires only fine-tuning. For *Merge* to work, somehow important brain

18   Johan Bolhuis, et al., "How Could Language Have Evolved?" *PLOS Biology*, 2014.

19   T. W. Deacon, "Human variability and the origins and evolution of Language," in *On Human Nature*, eds. Michel Tibayrenc and Francisco Ayala (San Diego: Elsevier, 2017), 557–64.

areas must be communicating with each other. But as to how *Merge* is implemented in the neural circuitry, we do not know, at least not at this point. Berwick and Chomsky discuss the little we do know at this point.[20] What they also mention is that a small genetic change in a growth factor for one of the essential fibers in the neural circuitry might very well be possible and essential for *Merge*—perhaps necessary, but not sufficient.

Chomsky himself minces no words: "It looks as if—given the time involved—there was a sudden 'great leap forward.' Some small genetic modification that somehow rewired the brain slightly [and] made this human capacity [for language] available.... Mutations take place in a person, not in a group.... It had to have happened in a single person."[21] This mutation in a single individual could have altered the wiring structure to link together language-related areas of the brain that are connected in human beings but remained isolated in non-human primates.[22]

Seen in this light, perhaps, or even most likely, some kind of "super-mutation" of gene duplicates was at the basis of a syntactically hierarchical language—although we don't really know. Let's not forget that mutations take place in a person, not in a group. But even if this mutation happened only to one individual, it could still be useful, because the outcome was not for communication but for thought. But that's where the discussion ends. The best we can say right now is that there is a deep language divide between the pre-human and human world. There is no way of telling whether this leap can be fully explained in terms of genes and mutations. Obviously, the faculty of language requires a body with the "right" features. The "right" mutations were *necessary* to achieve this. But whether those mutations were also *sufficient* is impossible to tell ahead of time. Those who think the language divide can eventually be completely explained by science must realize that this cannot be more than a conviction—more of a program than an achievement. There is no scientific way of corroborating such a conviction—it is worth as much as the opposite claim that a scientific explanation will never be fully possible.

The biologist Nicanor Pier Giorgio Austriaco, O.P. gives us a good summary: "Given the species universality of human language, the striking

20 Robert Berwick and Noam Chomsky, *Why Only Us*, 157–64.
21 Noam Chomsky, *The Science of Language: Interviews with James McGilvray*, ed. Noam Chomsky and James McGilvray (Cambridge: Cambridge University Press, 2012), 12.
22 Robert Berwick and Noam Chomsky, *Why Only Us*, 157–64.

architectural similarities among human languages, and the recent prov-
enance of human language, it is very likely that the capacity for human
language appeared only *once* in evolutionary history.... [So] all of us
today can say that we are direct descendants of the first speaking man,
because we too can speak."[23]

On the other hand, it must be admitted that there is much more than
this to the emergence of language. The most puzzling part of this event
was the emergence of the smallest atomic elements of language. Let's
call them "atoms of thought." Berwick and Chomsky are very aware
of the problem these elements pose: "Their origin is entirely obscure,
posing a very serious problem for the evolution of human cognitive
capacities, language in particular."[24] This shifts the problem from language to thought, from *Merge* to its *cognitive* elements, for language is
primarily a cognitive tool, not a communication tool. We could call this
an "ontological discontinuity." What we bump into here is the problem
of *rationality*—which takes us to the next chapter.

23   Nicanor Pier Giorgio Austriaco, "Defending Adam after Darwin: On the Origin of
Sapiens as a Natural Kind," *American Catholic Philosophical Quarterly*, 2018, 92 (2): 337–52.
24   Robert Berwick and Noam Chomsky, *Why Only Us*, 90.

7

The Rationality Divide

IF LANGUAGE IS INDEED A "SYSTEM OF THOUGHT," and if the smallest atomic elements of language are indeed "atoms of thought," then language is primarily an instrument of rationality rather than an instrument of communication. Let's see why.

In classical philosophy, human beings are called "rational animals." Why so? They are certainly animals, but then, animals with a full range of powers that are in addition directly ordered to thinking—rationality, that is. No wonder there is a strong connection between language and rationality, or between language and thought. Corrupted language may lead to corrupted thought, but also, vice-versa, corrupted thought may cause corrupted language.

What do we mean by rationality? Rationality is what distinguishes us from the non-human animals. It includes the power of *conceptual understanding*—the ability to understand the meanings of concepts—and the power of *conceptual reasoning*—the ability to judge the adequacy of these concepts and of the propositions that contain them. Let us discuss these two rational powers first before we address the question when and how they emerged in human history.

CONCEPTUAL UNDERSTANDING

A concept is the result of *abstraction* from what we have experienced through the senses. To be sure, all we know about the world does come through our physical senses, but this is then processed by the human intellect, which extracts from sensory experiences that which is intelligible in conceptual terms. For instance, we have seen several round objects and then we abstract from this the concept of "circle." This concept is abstract—in this case even highly abstract. It is very unlikely that we ever encounter a perfect circle in this world, which means we do not literally or physically "see" a circle. Besides, the concept of circle does not include any specific size, whereas the "circular" objects around us do. True, we can visualize a circle without imagining any specific size, but concepts have a universality that images can never possess. Therefore, the concept "circle" can be used for any specific circular object regardless of its size and its imperfections. That's what concepts can do for us.

In addition, a concept has an intricate web of connections with other concepts—in the case of a circle, for instance, with concepts such as "radius" and "diameter." Even a "simple" concept such as "green" or "greenness" has many connections to other concepts, which explains why we expect green objects to turn gray in twilight (the rod effect) and red when receding very quickly (the Doppler effect). As a result, concepts go far beyond what the senses provide—they transform "things" of the world into "objects" of knowledge, thus enabling us to see with our "mental eyes" what no physical eyes could ever see before. In short, concepts are cognitive tools.

All languages use *words*. What do words have to do with all of this? Most words have a *meaning*—they "mean" something to us and to other users of the same language. The meaning of a word is closely connected with concepts. The word "blood," for example, has a certain concept behind it—that is, the meaning of that word—which is connected with other concepts such as hemoglobin, blood cells, and the like. But there is another side to words. When we use words, we do so because words usually also have a *reference* (something they refer to). The word "blood," again, refers to the red fluid that circulates in our body or can be found in a blood stain. The difference between *meaning* and *reference* is quite crucial. The word "Venus," for example, can have very different meanings: for some it is about a planet, for some a Greek goddess, for others a horoscope—that is, the same word has very different meanings. On the other hand, the words "morning star" and "evening star" have each their own meaning, yet both refer to the same entity, the planet Venus—making them different conceptions of the very same thing.

Now it seems to be very attractive to think that a word is merely a label for what that word refers to. This could be called the "referentialist" doctrine, which claims that the meaning of a word lies in what it points out in the world, its reference. This view takes words as labels for concepts that refer to extra-mental, material objects in our world. But that view ignores the fact that the meaning and the reference of a word are not identical. The concept behind a certain word does not automatically or unambiguously refer to a particular physical object in the outside world. Without concepts, words don't refer to anything. Sometimes the relation of reference is hoped for (e.g., the "Higgs boson") or is imaginary (e.g., "centaur") or is stipulated (e.g., an irrational number such as π). Concepts don't *create* reality — they only try to *capture* it, and they may succeed or fail.

As Wittgenstein explained, pointing at things is not sufficient to define words and their concepts.[1] Of course, I can explain what the word "red" means by pointing at a red tulip. But that gesture is still very ambiguous. Perhaps someone else thinks "red" stands for a tulip, or for a flower, or whatever else might come to mind in connection with pointing at a red tulip. The word "red" is based on an abstraction — a concept, that is — of what we perceive. Without that concept it doesn't refer to anything. Besides, for many concepts there is nothing at which to point. To explain the concept "tomorrow," there is nothing to point at (other than on a calendar). And for the concept "pi" (π), there will never be anything we can point at. Consequently, an alleged proto-language not sophisticated enough to deal with the concepts of abstract thinking cannot really be called a language. Put differently, words are not mere labels assigned to things in the world around us.

Let's consider the example of a "gene." A gene may seem to be a physical concept with a clear reference: We can isolate genes, remove them, and replace them. But a "gene" is actually a mental concept — a concept that has an intricate web of related concepts behind it. It may not even "refer" to a specific, localized stretch of DNA, for genes can be overlapping (about 9% of human protein-coding genes overlap another such gene), so that a single stretch of DNA can be part of two or even three genes; sometimes the overlaps are partial; sometimes small protein-coding genes are fully embedded within much larger genes (e.g., the blood clotting factor VIII), so they form genes-within-genes. Consequently, the referentialist view is hard to defend here. What is much more essential to a concept is its meaning, so it can be used in thought and reasoning about genetics.

This is even more obvious in mathematics. Take for instance a mathematical concept such as π (pi). As the particle physicist Stephen Barr remarks, "One can have 4 cows, but one cannot have π cows; and one can have a 4-sided table, but not a π-sided table."[2] Why can π not simply

---

1 Ludwig Wittgenstein, *Philosophical Investigations*, §§ 258–77.

2 Stephen Barr, *Modern Physics and Ancient Faith* (Notre Dame, IN: Univ. of Notre Dame Press, 2015), 194.

be a property of material objects? Because there are no exactly circular objects in the physical world. Yet, although π does not have a *reference* in the physical world, we know the *meaning* of π. Something similar can be said about the concept "variable" as used in scientific and mathematical equations. It does not refer to anything particular in nature; it does not even appear in the equations themselves. And yet, it has a specific meaning.

It is very tempting, though, to get rid of the highly abstract and ethe-real notion of concepts by reducing them to something that is *neural*, but not mental. The famous example to defend this position is the following: When a frog sees a fly buzzing by, the frog's brain displays a certain pattern of neural firing. In a similar way, when we see a tree, there is a distinctive pattern of neural firing in our brain that is correlated with and caused by seeing a tree. This has led some to believe that thinking of a certain concept—say, a "fly" or a "tree"—is also and *only* a certain pattern of neural firing. However, the problem is that seeing a fly or a tree is a perception, whereas using a concept is a thought. Thoughts have meaning and content, which perceptions do not.

Concepts, thoughts, and beliefs are *about* something—they have con-tent and meaning. True, there is something like *perceptual* content in perception, but certainly not conceptual content. How can a group of material objects firing away have *conceptual* content? There is possibly perceptual content in a pattern of electrical nerve impulses, correlated with and caused by perception. However, conceptual content is not the same as perceptual content; it is about trees and flies in general, not about one particular tree or fly; it is universal, not particular; it is something mental, not neural; it goes beyond perception. That's why concepts have the power to help us understand and reason.

Another way of saying this is that we "see" particular things but we "reason" with universal concepts. Material things are always particu-lar—this circle, this square object, etc. I can see a *particular* circle, but I can think about *universal* circularity. If my thoughts about mathemat-ical circles were just a physical representation in the form of a neuronal firing pattern somewhere in the brain, those thoughts would at best be another particular material thing. But this necessarily means that those physical, material representations could not be universal. Reducing con-cepts to material entities—for instance, neuronal firing patterns in the brain—would make them something particular. Particular material things cannot qualify as universal. If the thought about circles were indeed a particular neuronal firing pattern, that pattern itself would have to be a circle as well—which is nonsense.

No wonder then that concepts play a central role in how we know the world. Thanks to concepts, we can see similarities that are not immediately visible and not directly tied to what we perceive. Everyone can see things falling, but to perceive "gravity" one needs the concept of gravity in order to "see" what no one had been able to see before Isaac Newton. The concept of gravity allows us to "see," for example, the similarity between the motion of the moon and the fall of an apple. But that is conceptual content, not perceptual. Animals can see things falling, but they don't see gravity. To "see" gravity one needs the concept of gravity, that is, to see what no one could see before Isaac Newton.

To take another example, biologists were not able to see the similarity in building blocks between animals and plants until the concept of a "cell" had been developed; neither could they see the similarity between leprosy and tuberculosis until the concept of "bacteria" had become available. Through concepts like these we are able to see similarities that would have eluded us if we didn't have those concepts. We do not really or directly perceive gravity, bacteria, cells, genes, circles, and the like, yet these concepts are essential to making the world more understandable and intelligible to us. Science, for instance, depends on them.

So the question arises: What are concepts such as π, circle, cell, gene, or species if they are not physical? The shortest answer is, they are non-material entities. Put differently, a concept—whether as simple as a "circle" or as complex as a "gene"—definitely goes beyond what the world shows us through our senses. We do not "see" genes but have come to hypothesize and conceptualize them. We do not even see circles, for a "circle" is a highly abstract, idealized concept.

No wonder, then, that concepts have an outlandish status. They are unlike anything else in this material world. Right in the middle of our comfortable, spacious, temporal, transient, and piecemeal world of material things, something immaterial pops up that we call "concepts." They are immaterial entities in a world of material stuff—they are immaterial, having no mass, no size, no color. In a world of material things there is only talk of being small, heavy, strong, and what have you—but not of being concrete or abstract, true or false, right or wrong. Suddenly we find ourselves in this strange, non-material world where things are not large or small, light or heavy, hard or soft. We can think about sizes and colors of things, but the thoughts themselves do not have sizes and colors. This makes the world of humans much richer than the world of non-human primates ever could be.

Although concepts are not material, humans could not live without them — either in daily life or in science. In their knowledge, humans do not deal with things directly, but after making a "detour" through concepts; they extrapolate from what is seen to what is unseen; they assign various interpretations from different perspectives to the physical things they see around them. They move from the world of sensible singulars, physical things, to the world of immaterial universals, concepts, and symbols. That is why the philosopher Ernst Cassirer suggested we call *Homo sapiens* a symbol-making animal [*animal symbolicum*]. Humans can create mental concepts that transform "things" of the world into "objects" of knowledge, as we said earlier. Concepts could be compared to a source of light, as the philosopher Edmund Husserl sees it: If a light beam hits a certain thing that is in darkness, this thing will be in the light, and yet it would not be inside the source of light.[3] Concepts are the "search lights" that change perceptions into observations, thus enabling humans to see with their "mental eyes" what no physical eyes could ever see before. In a sense, concepts illuminate what was previously in darkness.

It is to be expected that the word "concept" is also a concept in itself. It is the highly idealized "concept of a concept" that is at the heart of language, as we discovered in the previous chapter. If human language capacity is puzzling to us, how much more so must concepts be? Berwick and Chomsky are very honest when they admit, "In some completely unknown way, our ancestors developed human concepts."[4] Elsewhere they say, "The atomic elements [concepts] pose deep mysteries.... Their origin is entirely obscure, posing a very serious problem for the evolution of human cognitive capacities, language in particular."[5]

Now the question becomes more and more pressing: Where, then, do these concepts come from? Could they have come from the animal world?

---

3   Edmund Husserl, *The Crisis of European Sciences and Transcendental Phenomenology* (Evanston, IL: Northwestern University Press, 1970), 6.

4   Robert Berwick and Noam Chomsky, *Why Only Us*, 87.

5   Ibid., 90.

CONCEPTS IN THE ANIMAL WORLD?

The doctrine of gradualism makes it very tempting to assume that concepts must have emerged in evolution very gradually — with short steps and small mutations, one step at a time. Therefore, fans of this doctrine try to find instances of concept-like elements in the non-human animal world. But there are several reasons why concepts are not something we could ever possibly find among non-human animals.

Reason #1. Categories are not concepts.

Since concepts help us classify and categorize our experiences, gradualists will point out that animals are capable of classifying and categorizing as successfully as humans are: Animals are able to distinguish predators from prey, females from males, and so on. Obviously, the ability to classify is very advantageous to survival in an orderly world. As is to be expected, natural selection would favor such an ability. But does that mean animals too have concepts — the concept "predator" or "female," for instance?

It is indeed tempting to reason as follows: Concepts classify things; animals classify things; therefore, animals use concepts. But this is an invalid, false argument the conclusion of which does not logically and conclusively follow from its two premises. Obviously, in order to conceptualize, we often do need to categorize, but the reversed statement is not true: In order to categorize, we do not need to conceptualize. Concepts are our powerful human tools to see similarities even when they are not directly visible — for instance, the similarities between herbivores. But although animals too are able to see similarities, that doesn't mean they use concepts as well. Concepts are "ontologically" different from categories. How so?

Many concepts, such as "mammal" and "predator," do indeed classify and categorize, but we must keep in mind that classification and conceptualization are not identical. Chimpanzees, for example, classify colors the same way we do. They are also able to distinguish apples from bananas and then classify them together as fruits. But does this mean they use "fruit" as a *concept*? It is hard to see how this could be true. Classifying them as fruit is a very practical classification for very practical purposes. Let's take the example of buffalos: They do not identify a lion as a "predator" but as something to run away from — and they will do the same with other "predators." That's an inborn or learned response to an acute situation. The similarities animals see are not connected with concepts but with similar stimulus-response reactions. They learned, or were perhaps genetically programmed, to make certain associations. It's like children learning that things fall on the floor if they don't hold on to them—but they learn this without knowing about the concept of gravity.

Reason #2. Associations are not concepts.

Animals have the ability to learn to associate a certain input, a stimulus, with a certain output, a response. Stimulus-response associations are very common in the animal world. Humans have them too; but associations are not related to concepts. Associations are very common, but concepts make for a world of their own. Here is why.

Even chimpanzees—as "advanced" as they are on the way to humans-in-the making—are perfect examples of what Berwick and Chomsky call pure "associationist learners," who make direct connections between particular external stimuli and their signals.[6] They don't have mind-dependent concepts but only a series of mind-independent associations between objects in the external world and the signals taught them. These signals have a reference but no meaning, whereas concepts have meaning but not always a reference. When concepts do refer, they refer to the outside world, but from intricate perspectives. However, the signals we are talking about here only work when associated with a certain situation or emotional state. Even Jane Goodall, the closest observer of chimpanzees in the wild, had to come to the conclusion that for chimpanzees "the production of a sound in the *absence* of the appropriate emotional state seems to be an almost impossible task."[7] That must have been hard to admit for a gradualist like her.

What makes associations so different from concepts is the fact that concepts have no necessary link to a particular situation. We can talk about a "mammoth" without ever having seen one. We can talk about "Atlantis" without ever having been able to locate it. We can talk about "π" without ever being able to find an example of it in the real world. There are no associations involved.

6 Ibid., 146.
7 Jane Goodall, *The Chimpanzees of Gombe: Patterns of Behavior* (Boston: Belknap Press of the Harvard University Press, 1986), 125.

Reason #3. Labels are not concepts.

Words are sometimes considered "labels" for something that these words refer to. Numerals like "4" and "IV," for instance, are different labels we use to talk about the same abstract object: the concept of the number 4. The idea behind this is that concepts are the "meaning" of the word, and objects are the "reference" of the words. But as we found out, words in human language are not just "labels"; they come also with semantics and syntax.

True, animals may very well be able to use labels — as Sarah the chimp did with cards — but whether they can use concepts is very doubtful. The widely held "referentialist" doctrine that we spoke about earlier makes us believe that all concepts have a reference. This view takes words as labels for concepts that pick out extra-mental, material objects (their reference). However, concepts do not automatically or unambiguously refer to the outside world. The concept "boss" may refer to a particular person, but usually it can apply to different bosses in different situations. Besides, it may illuminate only a single aspect of what it refers to.

Apes, on the other hand, have never been seen using these labels other than as links associated with concrete objects in their surroundings. Nim was able to use the label "apple" but didn't have the human *concept* for "apple." As Laura Petitto observes, "Chimps, unlike humans, use such labels in a way that seems to rely heavily on some global notion of association. A chimp will use the same label *apple* to refer to the action of eating apples, the location where apples are kept, events and locations of objects other than apples that happened to be stored with an apple (the knife used to cut it)."[8] In other words, the label *apple* was used to refer to *anything* associated with apples. But that doesn't make the label a concept.

Reason #4. Sensations are not concepts.

Sensations, or sense-impressions, are the data we receive through our senses. Animals have sensations too, of course. Most sensations come in through the eyes, especially so for vision-oriented animals; they make for visual sensations received from the retina. But images are not concepts.

---

8   Laura Anne Petitto, "How the brain begets language," in *The Chomsky Reader*, ed. James McGilvray (Cambridge: Cambridge University Press 2005), 85–101.

No matter how many pictures of cells we have seen through the microscope, the concept "cell" does not automatically arise from these images or their sensations. No matter how many cows, horses, and goats we have seen on our retinas, the concept of "herbivore" does not spontaneously emerge from these sensations. It took the biologist Claude Bernard a while to realize that the urine of animals like horses and rabbits is turbid and alkaline because they are herbivores, whereas carnivores such as cats and dogs have clear and acid urine. In making this connection, Bernard went far beyond the individual cases he had seen repeatedly.

Sensations, images, and pictures are very ambiguous, open to various interpretations. For sensations to become observations, we need concepts to make sense of these sensations, to make them more specific, and to make them intelligible for the human mind. What scientists, for instance, report as the result of an experiment is not a recital of sensations they had. It is rather an *interpretation* of these sensations, by transposing them into the abstract, symbolic world of concepts by means of abstraction. Abstraction allows us to focus on certain similarities while leaving out the dissimilarities. Karl Popper used to say that the command "Observe!" does not make any sense, since no one would know what to observe. His point is that scientific theories just do not and cannot spontaneously emerge from sensations; they do not spontaneously pick out similarities. We do not "have" observations—like we have sensorial experiences—but we "make" observations.

Undoubtedly, animals are also able to see similarities by mere sensation; that's how they can identify and recognize and categorize food, predators, and mating partners. So they may even be able to recognize a circle when presented with a circular object, which could be considered a primitive form of abstraction. However, their seeming act of abstraction is closely tied to what the philosopher Mortimer Adler calls a "perceptual act."[9] In experiments, dogs, for instance, can only recognize a circle when presented with an actual circular object. So this kind of abstraction could rightly be called "*perceptual* abstraction" because it is closely tied to actual perception.

Humans, on the other hand, can also see similarities through

---

9 Mortimer J. Adler, *Intellect: Mind over Matter* (New York: Macmillan, 1990), chapter 4.

concepts—that is, apart from any perceived object. Only human beings can think about a circle or about circularity in general, apart from any specific perceived object. That's what we do in geometry, for example. We could call this, with Adler, "*conceptual* abstraction" to distinguish it from "perceptual abstraction." Adler maintains that there is no scientific evidence that any animal other than human beings can understand universals, detached from particular sensations.

Reason #5. Signals are not concepts.

Words can either be used as *signals*—to refer to physical entities—or they can be used as *symbols*—to refer to mental entities (especially to abstract concepts). Take for instance a word such as "poison": it can be either a symbol referring to a concept that explains its nature and working, or it can be a signal or a label on the bottle that alerts us not to take that stuff.

This difference also separates the world of animals from the world of humans. Humans can use both signals and symbols, but animals can use only signals. Animals treat everything in their surroundings as signals that call for a direct response (association), but they cannot use concepts to ponder realities beyond their needs for food and sex. They act mainly by "instinct," not by conceptualization. Even the sign language people with hearing impairment use is a real language, so it doesn't use signals but signs that refer to concepts.

Many animals are able to communicate with each other through signals—just listen to birds on an early spring morning. However, signals are not symbols, let alone concepts; those birds are having an exchange of signals, but not an exchange of ideas. Signals refer directly to a specific thing or situation, whereas concepts usually do not. When prey animals flee away from a predator, they are not reading the predator's mind but are processing certain signals. Animals live in a world of signals, while humans live in a world of signs and symbols.

A prey animal, for instance, can only take a "predator" as a *signal* to flee or attack, but not as a *symbol* with various interpretations—for example, as an animal in need of food, as an animal who is born to prey, as an animal brought up that way, as an addicted killer, as a pre-programmed killer, as a member of a larger conspiracy, as an inevitable part of life, or as a part of nature that needs to be preserved. Humans can come up with such different interpretations, but animals cannot, for animals see only signals that call for an immediate response (true, we do too sometimes!).

Animals do not have this capacity. When I warn a dog by pointing my finger to an approaching car, the dog just looks at my finger—and may even lick it—but it does not get what my finger refers to; it cannot make various interpretations. For animals, my pointing finger is just what it is, a finger (although they can be trained to associate this with something else as a signal). My finger cannot just direct the animal's attention to something beyond itself, for that requires interpretation and thought.

Thanks to concepts we can *interpret* what we perceive. For example, when we describe what we see in the sky as "Those are moving dots," we use very vague concepts—"dot" and "move"—and therefore very little interpretation. When we say, instead, "Those are flying birds," we use more explicit concepts—"bird" and "fly"—and therefore implement a more specific interpretation. And when we say, "Those are migrating geese," we inflate our interpretation even further with more explicit concepts—"geese" and "migrate." Obviously, animals do not—and arguably cannot—do this. If they spot a hawk in the sky, they are not identifying the bird with a concept ("hawk") but merely identifying it as a signal of imminent danger that may require direct action. They don't study birds in the sky as ornithologists do.

Reason #6. Commands are not concepts.

Animals can and do communicate with each other through various kinds of signals—aggressive displays, courtship displays, facial expressions, and so much more. Many signals in the animal world work like warning signs, or even commands. They include commands that warn other animals to back off. But they can also be commands telling other animals to hide from imminent danger. Vervet monkeys, for instance, have been shown to use different signals to warn others of different types of enemies.[10] But commands like these are very different from concepts. Those monkeys use their different signals only when a particular enemy is around; they do not use them to just "ponder" and "chat" about a particular enemy in general during their spare time. Signals are situation-specific.

10   Robert M. Seyfarth, et al., "Vervet monkey alarm calls: Semantic communication in a free-ranging primate," *Animal Behaviour*, vol. 28, issue 4, 1980: 1070–1109.

Let's explain this further with a simple example. When we train a dog to associate a command such as "The boss!" with its real boss, then the dog has been conditioned to respond to such a command by looking for the real boss. The command has become a *signal*. Human beings certainly share this feature with animals; one employee yelling "The boss!" is actually signaling all others present to start looking extremely busy. That's what signals do; they depend on the actual presence of the "real thing" — for instance, due to associative conditioning through training. The dog has a physical image of its own boss, but it has no mental concept of what a "boss" — any boss, for that matter — is like. Humans, on the other hand, can use the word "boss" also as a *symbol* — a mental concept of "any boss" in general. They often use that word to talk about what their own boss is like, or should be like — preferably only when their physical boss is not actually present. Signals call for direct action, whereas concepts do not.

Consequently, it does not matter whether you train a dog with a command like "Here!" or a command like "Hector!" Dogs react the same way, not realizing the latter command refers to themselves. For animals, either command works through association, but only humans know they are fundamentally different. Animals cannot make this distinction, so they treat everything in their surroundings as signals that call for an immediate, direct, and definite response. Animals are "born positivists" — they take everything as a signal at face value. They can handle signals but not symbols. Your pets, for instance, do not chat about or meditate on what their boss, you, is like — they are not pondering creatures.

Humans, on the other hand, can deal with things after making a "detour" through symbols and concepts; they extrapolate from what is seen to what is unseen; they can assign various conceptual interpretations to the things they see. They move from the world of sensible singulars (things, situations, and events) to the world of immaterial universals (concepts, symbols, and facts). That is the reason why *Homo sapiens* has also been called an *animal symbolicum*.

Let's come to a conclusion: Could humans have inherited their faculty of using concepts and symbols from the animal world? After what we have seen, that is a position hard, if not impossible, to defend. We have the same problem here as we had with the origin of language use. The power of using language and the power of conceptual understanding are both uniquely human, and therefore could not be inherited from the animal world. Does that mean neither one can be explained by genetic changes? Not necessarily, but assuming they can is more of a program than an achievement. On the one hand, Berwick and Chomsky realized that "the origin of mind-dependent word-like elements remains a big mystery — for everyone, us included."[11] On the other hand, they did speculate that at least some of these elements existed prior to *Merge*, since otherwise there would be nothing for *Merge* to work on.[12]

CONCEPTUAL REASONING

We said at the beginning of this chapter that rationality includes not only the power of *conceptual understanding* — the ability to understand the meanings of concepts — but also the power of *conceptual reasoning* — the ability to judge the adequacy of these concepts and of the propositions that contain them. So reasoning is indeed based on concepts but also moves beyond them by using them in propositions. Both concepts and propositions are highly abstract objects. The concept "snow" is an abstract notion for what all snow has in common (Eskimos have dozens of different words for different kinds of snow). When we use this concept in a proposition such as "snow is white," we can use different languages at different times and at different places to express the same proposition — namely, the proposition that snow is white. It is an abstract thought, not a concrete sentence spoken or written in a certain language at a certain time and place. That's why all people can mean the same thing when they say, in whatever language, that snow is white.

Rationality is our capacity and faculty to make propositions, judgements, and decisions (which does not mean, of course, we always think rationally!). In fact, it is the faculty of rationality that gives us access to the world of truths and untruths — a world beyond our control. Rationality is our capacity for abstract thinking and having reasons for our thoughts, thus giving us access to the "unseen" world of thoughts, laws, and truths. Rationality allows us to gain knowledge about the world through the power of abstract concepts and mental reasoning, thus giving us an immaterial sense for what is true and what is false. Weighing evidence and coming to a conclusion are rational activities *par excellence.*

11   Robert Berwick and Noam Chomsky, *Why Only Us*, 149.
12   Ibid.

Reasoning leads us from one idea to a related idea; it is a matter of pondering realities beyond those which we experience through our physical senses. Philosophical giants such as Aristotle and Thomas Aquinas would put it this way: All we know about the world comes through our physical senses but is then processed by the immaterial intellect that extracts from sensory experiences that which is *intelligible*. Well, it is the rationality of our intellect that makes the world intelligible and understandable; that gives us the power to comprehend the Universe through reasoning. It is the mind's faculty of rationality that gives us access to the laws of nature and the structure of this Universe. Laws of nature have to be discovered, not invented; they are not just mental creations but are anchored in reality. Just think of all those scientific explanations that are based on scientific laws of nature — they explain material things and events, and yet they use non-material laws of nature to do so.

Truth claims are immaterial assessments regarding the immaterial world of *facts*. Facts are considered by many as the "rock-solid" or "hard-core" realities we bump into. However, as a matter of fact, facts are not physical objects, but rather intellectual, rational entities. They are our interpretations of reality — of things, situations, and events around us — thus making the world intelligible for us. In other words, there are no facts without human thoughts, concepts, statements, and interpretations. We literally see material things, situations, and events, but we cannot literally see immaterial facts. It is a fact, for instance, that Darwin never met Mendel; obviously, there is no event we can point at. Facts are our mental interpretations of things, situations, and events. This does not mean, though, that facts can change — facts are facts, always and everywhere. Although facts cannot change, sometimes we declare something a fact, which turns out on further investigation not to be a fact. Whereas animals live in a world of objects, humans live also in a world of facts.

Reasoning works with arguments. Arguments usually lead to conclusions based on some premises. Traditionally, arguments are divided into two different types, deductive and inductive. Although every argument involves the claim that its premises provide evidence for the truth of its conclusion, only a deductive argument involves the claim that its premises provide *conclusive* evidence. In deductive reasoning, premises and conclusion are so related that it is absolutely impossible for the premises to be true unless the conclusion is true as well. In other words, if the premises are true, then the conclusion *must* be true.

Here is a simple example of a valid deduction: (1) All humans are mortal (premise); (2) I am a human (premise); (C) I am mortal (conclusion follows from premises 1 & 2). This argument seems to be valid, since the truth of the premises would guarantee the truth of the conclusion. And it also seems to be truthful, since, in addition, the premises do seem to be true as well.

Deductive arguments are most common in mathematics, but they are often also used outside the domain of mathematics. An example of this would be the following: Since tests proved that it took at least 2.3 seconds to operate the bolt action on Lee Oswald's rifle, Oswald could not have fired three times to hit President John Kennedy twice and Texas Governor John Connally once in 5.6 seconds or less, for that would take 3 x 2.3 = 6.9 seconds. This seems to be a valid conclusion of the deductive type. However, this argument uses an extra, hidden premise that can easily be overlooked, namely that in the time it takes to fire three shots, it is only necessary to operate the bolt twice. So then the conclusion is no longer necessarily true. It turns out to be an example of faulty reasoning.

Hidden premises or assumptions more often than not put the conclusion of a deductive argument in jeopardy. Columbus, for instance, reasoned that the earth must be round in the following deductive way: As a ship sails away from shore, the upper portions of it remain visible to a watcher on land long after its lower parts have disappeared from view—so the earth must be round. But again, there is a hidden premise involved, which states that light rays follow a rectilinear path. If they followed a curved path, concave upwards, then we would still see the same happening to the ship, even on a flat earth! So this argument only qualifies as deductive, leading to a conclusive conclusion, if we specify all the necessary premises involved and accept them as true. No wonder some of Columbus's sailors were still afraid to fall off the edge of the world.

As a matter of fact, each time we engage in a discussion or dispute, we use reasoning to defend our position or to explain why we disagree with the position of others. Each time we look for any kind of explanation we are in search of some form of reasoning. Aristotle once came up with the following line of reasoning about lunar eclipses: Since it is the interposition of the earth that causes the eclipse, the form of this line will be caused by the form of the earth's surface, which is therefore spherical—and not flat.[13] Thanks to reasoning, he was far ahead of his time without any fancy research. That's just another simple example of what reasoning can do for us.

An *inductive* argument, on the other hand, involves the claim that its premises provide *some* evidence for the truth of its conclusion—but no conclusive evidence. The truth of the conclusion does not follow necessarily from the truth of its premises. In other words, if the premises are true, then the conclusion *may* be true. Thus the conclusion is only probable, or probably true. Very often an inductive argument is an argument by analogy. For instance, we infer that we will enjoy reading a book by a certain author on the basis of having read and enjoyed other books by that author. We may find however that our favorite author's latest book is actually a bummer.

Yet some inductive arguments by *analogy* are more cogent than others. The problem is that we need to find out first which similarities, or analogies, between certain cases are *relevant*. Imagine, you love pets and have repeatedly noticed that your dog and your cat have clear urine. You might think being a pet animal is the similarity that causes their urine to be clear. But that would mean your pet rabbit and your pet hamster should also produce clear urine, until you discover they actually have cloudy urine. What causes this difference in urine? You might come up with various kinds of explanations of why their urine differs. Well, it turns out that once you know that there are carnivores and herbivores, you have found the correct explanation for a difference in their urine. This step could not have been made without two new concepts: "herbivore" and "carnivore." There are those concepts again!

It is the mind's faculty of rationality that gives us access to the laws of nature and the structure of this Universe. Right in the middle of our comfortable, spacious, temporal, transient, and piecemeal world of material things, something pops up that we call "laws of nature"—physical laws, Mendelian laws, mathematical laws, and even moral laws. Unlike all material things surrounding us, laws do not have any of the features that apply to the material world—that's right, none. Yet, when scientists or engineers violate these laws, they get themselves into real trouble. A

13 Aristotle, *On the Heavens*, book II, chapter 14.

bridge that has been designed according to the right laws can stand firm, whereas another bridge collapses because its engineers erred in their calculations—perhaps they had the wrong laws in mind, or at least the wrong thoughts. The construction of a bridge would never depend on the right laws and the right thoughts, if those laws and thoughts were only creations of the human mind. It would not make sense to say that competent engineers have better mental habits than their inept colleagues. There must be more to it.

The question arises again: Can reasoning be traced back to the animal world? It is unlikely, because the use of concepts seemed to have popped up from nowhere in evolution—and reasoning assumes concepts. But let's not jump to conclusions yet.

REASONING IN THE ANIMAL WORLD?

Gradualists—mindful of Darwin's goal "to show that there is no fundamental difference between man and the higher mammals in their mental faculties"[14]—would maintain that there must be "something" like reasoning in the animal world. If animals were to do "stupid" things they would not survive long. That sounds like an unbeatable argument. But perhaps we should introduce an important distinction first: the distinction between intelligence and intellect.

There is an important difference between these two. Even when animals use their intelligence, they can do so without intellect. How come? Intelligence works mainly with stimuli, signals, and sense impressions—that's why even robots can act in an "intelligent" way (artificial intelligence). Intellect, on the other hand, is very different from intelligence and works with concepts, symbols, and reasoning. Rationality is not a matter of intelligence but of intellect. Whereas intelligence can be graded on an IQ scale, intellect cannot. One can have more or less intelligence but one cannot have more or less intellect (one can be more or less intellectual, though). The intellect allows us to *understand* the meanings of concepts and of the propositions that contain them, and it allows us to *judge* the adequacy of these concepts and the truth of these propositions.

Gradualists are right in one respect: When it comes to intelligence, humans do not seem to be exceptional creatures. Intelligence is a rather

14 Charles Darwin, *The Descent of Man*, 1871/1896, 66.

manifold entity; it has many faces — social, practical, formal, spatial intelligence, and so on. That explains why many animals do show some form of intelligence in their behavior, because intelligence is a brain feature and as such an important tool in survival. Intelligence is a matter of properly processing sense-data — something even a robot can do by "cleverly" processing sounds, images, stimuli, signals, and the like. Even a simple pocket calculator is able to "cleverly" manipulate mathematical symbols. Not surprisingly, then, animals show various forms of intelligence: We find spatial intelligence in pigeons and bats, social intelligence in wolves and monkeys, formal intelligence in apes and dolphins, practical intelligence in rats and ravens, to name just a few. When some scientists assign *cognitive* capacities to non-human animals, they are basically referring to what most people take as *intelligence*. Birds such as crows, ravens, and blue jays are very smart — or call it intelligent, if you want — in certain cognitive domains, even though they must achieve this without concepts. But using the term "cognitive" is actually deceiving since there is no cognition involved that requires concepts and reasoning.

So animals may indeed have different degrees of intelligence, although arguably not of the high standard found in humans. Why is it that animal intelligence so often disappoints us? Could it be the fact that they lack rationality and intellect, in addition to intelligence? As we said already, rationality, which comes with intellect, is different from intelligence; intelligence may provide some helpful survival tools, but ultimately it's our intellect that allows us to understand and explain the world we live in, by the use of mental concepts. Like intelligence, intellect also uses sense-data, but unlike intelligence, it changes perception into cognition by means of mental concepts and logical reasoning, which makes sensorial experiences intelligible for the human mind. When a dog avoids another dog that it was bitten by before, it does so because a material cause — a bite — makes the dog associate the bite with a certain dog. It is a result of conditioning, not reasoning. Animals drink because they are thirsty — which is not a mental reason but a material cause. Human beings, on the other hand, can drink for all kinds of reasons other than quenching their thirst.

How did it happen that humans — arguably unlike non-human animals — became able to think consistently and reasonably? A supermutation perhaps? If that were the case, natural selection could explain how our thoughts came to correspond in some way to reality, for "not thinking straight" would be a disadvantage in the struggle for survival. That argument certainly sounds attractive but it is not really adequate, for several reasons.

First, humans are able to do various things in reasoning that have nothing to do with survival in the wild. The human intellect gives us the power to "understand" a wide variety of things, far beyond mere survival. Knowing that the earth is not flat doesn't help us in the struggle for survival. How could one possibly explain that natural selection gave us a genetic program that was vastly more sophisticated than was required for survival?

Second, success in survival does not equate to truth: True explanations may indeed be successful and useful, but successful or useful explanations are not necessarily true. Ptolemy's geocentrism, for instance, was successful and useful in navigation, but not true. As a matter of fact, truth is often not particularly useful, whereas many an illusion proves useful.

Third, natural selection cannot explain that we know certain thoughts as being "necessarily" true, such as $2 \times 2 = 4$. This kind of certainty and necessity is beyond the trial and error strategy of natural selection. Natural selection cannot possibly produce the certainty that 11 is a prime number. So the question is how natural selection could produce such knowledge — plus the knowledge that the laws of logic and mathematics must be without any exceptions.

Fourth, the claim that our knowledge of the world is merely the outcome of natural selection can only be made by using highly sophisticated and abstract concepts such as selection, genes, mutations, and so on. And those are exactly the concepts that are assumed to be the product of natural selection, which basically leads us into a vicious circle — explaining concepts with the very concepts that need to be explained. In other words, there are serious reasons why this endeavor is bound to fail, making it hard to believe that the human intellect is a product of survival.

Though the intellect may not always enable us to survive better, it does enable us to think properly. Take the principle of non-contradiction. It is usually stated that a thing cannot "be" and "not be" at the same time in the same circumstances. If two things were absolutely identical in everything, including their existence, they would be the same being and the question of difference could never arise. So how do we know then that this principle is true? Not because someone told us so. Not because our genes told us so. Not because natural selection promoted it. It is because we cannot think it to be "not true" without at the same time affirming that it is true. With this principle of non-contradiction we can begin to distinguish and separate things. We can begin to put some order into things: this thing is not that thing; this thing is like that thing but is different too. It is the intellect that makes this kind of reasoning possible.

Whereas reasoning is pondering realities beyond that which is experienced through the senses, animals, in contrast, seem to live their lives entirely in the present, without having any thoughts about the past or the future — perhaps memories, but not thoughts, symbols, or interpretations. If pets have a pedigree, it is thanks to their owners; if they have birthdays, wish lists, appointments, or schedules, it is because their owners create those; and if they have graves, those were dug by their owners as well. Cats or dogs can act rather "smart" and intelligent, but they have never come up with the thought of going to the pet store themselves and buying their own food, let alone of starting their own pet store.

Yet, gradualists keep claiming that animals must have some form of rationality, because the difference between animals and humans has got to be one of degree, not kind. So they have come up with experiments to prove their point. However, most experiments that pretend to demonstrate *reasoning* in animals have the following problem: The test results they take as evidence for conditional reasoning (if-A-then-X, or if-B-then-Y) can equally well, or even better, be explained by associative conditioning (A-and-X, or B-and-Y). Hence, opting for the assumption of reasoning is somewhat arbitrary if we don't have additional, compelling reasons for making this choice — yes, reasons again!

Let's study the following classical experiment as a test case.[15] Rats — pretty smart animals with intelligence — were trained to expect food to drop into a little tray after a certain sound. Next those rats went through a second training in which they found food in the tray, without a sound, but with a vomitive in it. After these two training sessions, we might expect that a sound would make them go to the tray (as a result of the 1st training), but then they would not touch the food (as a result

---

15   P.C. Holland and R.A. Rescorla, "Second order conditioning with food unconditioned stimulus," *Journal of Comparative and Physiological Psychology*, 1975, 88: 459–67.

of the 2nd training). In reality, they gradually stopped going to the tray! In a control experiment, the 2nd training was modified: Those rats would only find a vomitive in the tray. The rats that went through these two training sessions kept going to the tray after each sound, although there was no food in it!

The scientists who did this experiment went happily for the first explanation, conditional *reasoning*. They claimed the rats in the first group made logical inferences based on mental concepts. The rats supposedly had two expectations: #1 "sound causes food" (if A then B) and #2 "food causes vomiting" (if B then C). And from this, these smart rats deduced the following thought: #3 "sound causes vomiting" (if A then C). In other words, "food" (B) had become a mental concept, so it could act as a logical link between a sound (A) and vomiting (C), making vomiting a reason to stop going to the tray. The proof for this conclusion is supposed to be in the control group, for that group kept going to the tray, showing us that the first group didn't get tired of going for nothing but must have had a "reason" to stop going.

However, we should ask these scientists: Why not go for the other option, associative *conditioning*? The rats in the first group learned to associate sound with food (A-and-B) and food with vomiting (B-and-C), so they didn't touch the food in step 2, hence were not rewarded, and thus gradually stopped going to the tray. The second group also learned to make the same first association, "sound with food" (A-and-B), but the second association was "no-sound with vomiting" (not-A and C), so they kept going to the tray—which doesn't seem to be very rational, by the way.

It could perhaps be argued that either explanation is valid, but if one does want to go for conditional reasoning, which has much wider ramifications, one would need additional arguments, which seem to be lacking. So then the only reason to vote for reasoning over conditioning would be ideological—gradualism, that is. Obviously, it remains a perpetual temptation to interpret animal behavior in terms of rationality, which is done by equating motives to reasons. Animals often do have motives—for instance an internal drive to cross a river or a drive to search for food or a drive to find a mate. But this doesn't necessarily imply that they make reasoned decisions as well. Animals can even react to the drives of other animals, but that doesn't mean they can "read each other's mind." All we can say is that they learned to associate certain behaviors with certain outcomes—but let's not confuse this with rationality.

Let's get to some animal ancestors that are much closer to us than rats. These "close relatives" have indeed shown some striking cases of intelligence. But let's not go overboard. More than a century ago, Edward Thorndike, the pioneer of American experimental psychology, complained bitterly about all the books of his day (published after Darwin) which, he said, gave not "a psychology, but rather a eulogy of animals. They have all been about animal intelligence, never about animal stupidity."[16] He was right, there's certainly a lot of "stupidity" in the animal world when it comes to rationality, or even formal intelligence. Just think of the following cases:

- The cat that was placed in a closed box and then discovered after many trials how to pull a cord in order to open an escape hole, still keeps pulling the cord before crawling out, even when the escape hole is already open.
- Or the cat that had finally learned to escape by pulling a cord, but when subsequently placed in the same box without the cord, goes through the motions of pawing the air where the cord had been before.
- The elephants that had learned to lift a lid to retrieve food from a bucket, didn't "get it" when the lid was placed alongside the bucket while the food was simultaneously placed inside the bucket; all trained elephants continued to toss the lid before retrieving the reward, raising the possibility that they have no understanding of this simple causal relationship.
- The chimpanzee that has learned to invoke help from a trainer but does not know which trainer to call upon if one of them has a bucket over his head while the other has not.
- The chimpanzee that cannot learn to secure a larger reward by pointing to a smaller one. The immediate urge for food just overrules any intelligent alternatives.
- And my favorite one: Some tribes in Africa make a split in a coconut, wide enough to get a hand through, and fill it

---

16 Edward Thorndike, *Animal Intelligence* (New York: Macmillan, 1911), 22.

with candy as a way to catch monkeys who put their hand inside the coconut to grab the candy, but can't get rid of the heavy coconut because they keep their fists tightly closed around the candy.

A much more recent voice puts it this way: "I diligently recorded and reported the hundreds of trials it took the chimpanzees to learn what, in truth, seemed like a rather basic problem—a problem that should have been well within the abilities of the chimpanzee genius.... I began to understand that many of my most cherished beliefs about chimpanzees were based on faith, not evidence."[17]

More generally it has been claimed that, depending on their intelligence, animals occasionally show some capability of seeing "physical" connections between things. Because of this, animals are presumably able to perform trial-and-error activities. Chimps, for instance, have been able to "see" that a stick can help them reach food by making such a "physical" connection when they see a stick near food. However, when the food and the stick are kept far apart, they cannot make the connection—even if they were able to do so before. Apparently the stick is not a mental concept but rather a physical signal that can invoke an intelligent reaction; a chimp has a physical image of a stick in relationship to food, but it has no mental concept of what a stick is and can do in various situations. Why not? Animals are fully enveloped in their surroundings, but they are not subjects capable of creating mental concepts transcending the immediate situation; animals are caught up in a network of physical connections, unable to transcend the current situation with the mental power of abstract concepts and reasoning. As said earlier, *perceptual* abstraction is different from *conceptual* abstraction. This explains why intelligence differs fundamentally from the rationality of the intellect.

In short, animals may be more or less intelligent, but they are not rational beings. Pets, for instance, can even be smarter than their owners when they play a whole repertoire of tricks on their owner's emotions—but that is a matter of intelligence at best, not intellect. Animals do have the capacity to sense and remember things, but they lack understanding in

---

17   D.J. Povinelli, *Folk Physics for Apes: The Chimpanzee's Theory of How the World Works* (Oxford: Oxford University Press, 2000), x.

the sense of asking questions, formulating concepts, framing propositions, and drawing conclusions. They show no signs of abstract reasoning or having reasons for their "thoughts" (if they have any); they do not think in terms of true and false; they do not think in terms of cause-and-effect; they do not think with "if-and-only-if" statements. Instead, they are "moved" by motives, drives, instincts, emotions, stimuli, and training, but not by reasons or mental concepts. In other words, animals do not have an intellect endowed with the capacity for rationality — regardless of their intelligence. Obviously, so we found out earlier, they don't have the faculty of language either, which is an important tool for thought as we found out also.

Nevertheless, some gradualists have even assigned "mind reading" to animals. They report, for instance, how a subordinate chimp stayed away from hidden food when she had watched how a dominant chimp saw researchers hide the food.[18] The authors of this study argued that this was because the subordinate chimp *reasoned* about what the dominant chimp had seen and what he would do next. Combined with results from other experiments, they concluded that chimps can understand both the goals and intentions of others as well as the perception and knowledge of others, and that they can predict the action that will result. Chimps have even been awarded with a "theory of mind." However, this reeks of extreme anthropomorphism. We might arguably speak of unscientific inferences based on our own human experiences. The subordinate chimp doesn't have to know the "mind" of the dominant chimp, if the chimp has one; all she has to do is to avoid interfering with the dominant chimp.[19] The dominant chimp has in fact become a signal for others: Be careful with that one!

Since we are masters of anthropomorphism, we tend to think that animals have got to be like humans, even with regard to rationality. But what a disparity there is between them and us! Only humans are conscious of time; they can study the past, recognize the present, and anticipate the future; they even desire to transcend time, thinking about living forever. Only humans wonder "what caused or will cause what, and

18 Brian Hare, et al., "Do chimpanzees know what conspecifics know?" *Animal Behaviour*, 2001, 61: 139–51.

19 Joseph Call and Michael Tomasello, "Does the chimpanzee have a theory of mind? 30 years later," *Trends in Cognitive Sciences*, 2008, vol. 12, issue 5: 187–92.

why?"—thanks to the concept of causality. Only human beings have inquisitive minds asking questions such as "Where do we come from?" and "Why are we here?" Only humans have the capacity to be scholars and scientists; they can even study animals, whereas animals can only watch humans but never study them. Human beings are always in search of some kind of worldview or explanation of life—which certainly goes far beyond their need for food and sex. From early childhood on, they never seem to tire of asking "Why?" In short, human beings are inquisitive, questioning beings; they are driven by rationality, which gives them the capacity to make rational decisions.

Of course, it is hard to prove that there are no genes for rationality, because those genes may be eluding us or are still waiting to be discovered. When we say there are no black swans, for instance, we may not have searched long enough. Instead, it may be easier to prove that there is no rationality in the pre-human animal world. But again, that may also be a matter of not having searched long enough. So we probably have a much better chance to prove *a priori* that there cannot possibly be any rationality in the pre-human world on logical and philosophical grounds. That's what we tried to do in the previous paragraphs.

Let's bring this discussion to a close. Are we the only thinkers on planet earth? When it comes to intelligence, humans may not be the only thinkers here on earth; they are arguably far the greatest, but humanity is not measured by intelligence, as intelligence is just a matter of sophisticated brain wiring. More intelligence doesn't make us more human; it is the rationality of the intellect that makes us human. Animals may have intelligence to a greater or lesser extent, but they have no intellect to a greater or lesser extent. There is a deep divide here. It is the human mind that has this unique ability to think under guidance of the light of reason. There is no evidence that the divide can be completely explained by gene mutations and natural selection. Those who think the divide can eventually be fully reduced to mutations must realize this cannot be more than a conviction, worth as much as the opposite claim that a scientific explanation will never be fully possible.

So we must come to the following conclusion: No animals seem to pass the "rationality test." There is a deep divide between them and us in terms of rationality. It is hard to believe that reasoning, including the laws of logic, came from the animal world. If they did, Chesterton would like to ask those who think so, "Why should not good logic be as misleading as bad logic? They are both movements in the brain of a bewildered ape?"[20] If that were true, good logic or math would be as misleading as bad logic or math. But there is something about good logic and good math and good science that lifts them far above what evolution has produced.

This may remind us of Ludwig Wittgenstein's remark: "We say a dog is afraid his master will beat him; but not, he is afraid his master will beat him tomorrow."[21] Why not? Because a concept like "tomorrow" (or "gravity" or "force" or "causality") is part of *conceptual* abstraction, rather than perceptual abstraction. For animals there is no "tomorrow," because that requires the concept of "tomorrow." Besides, language is necessary to talk about concepts that stand for inferred, abstract properties (e.g., gravity, force, tomorrow). Deprived of these concepts, non-human animals must work exclusively with what is directly observable.

ONLY HUMANS DEAL WITH FACTS

How could facts have only appeared when the first humans appeared? That goes against good judgment, doesn't it? Common sense tells us differently: Facts have always been there, no matter whether humans were around or not. Facts are supposed to be the "rock-solid" or "hard-core" realities that all animals, both human and non-human, have to deal with. All animals who deny the facts or go against them will be ruthlessly eliminated by natural selection. That seems to be a fact in itself and could not have changed when humanity emerged in evolution.

The idea that facts have always been there could easily be misunderstood. On the one hand, it's true that facts have always existed, with or without the presence of human beings. The fact that Planet Earth is spherical has always been a fact. Facts are facts, always and everywhere. In that sense, there have always been facts, for facts never change. Facts are supposed to be detached from time and space; they are true regardless of who you are and when and where you live; they appear as objective, absolute, and universal. In that specific sense, facts have always been there, no matter whether humans were around or not. Facts are facts—whether

---

20 G. K. Chesterton, *Orthodoxy*, 33.

21 Ludwig Wittgenstein, *The Blue and Brown Books* (New York: Harper Torchbooks, 1958), 166.

we like them or not, whether we know them or not, whether we have discovered them or not. Facts are facts forever, so to speak.

On the other hand, facts do not exist the way material things around us exist, because facts are not material things—they are mental objects. Things and events are "physical" parts of our world, perhaps even rock-solid, but facts are "mental" creations which cannot be physical or rock-solid—which does not mean, by the way, that facts only exist in the mind. A "fact" is not like a "thing" you can kick. We are indeed surrounded by concrete things, events, situations, and processes, but facts do not belong to this list—they are of a different nature. Animals do live in a world of things, situations, events, and processes, but they don't live in a world of facts. Facts cannot exist without concepts. Actually, a "fact" is another *concept* of its own, and concepts are a unique part of human beings, as we discovered earlier in this chapter. So, in that particular sense, facts are tied to human beings. At the moment we start thinking and talking about them, facts seem to come back into space and time, because thinking requires thoughts (rationality), and talking requires statements (language). And that's what makes the situation so complicated, because now we end up with at least four rather disparate elements: events, thoughts, statements, and facts.[22] How are they related then? Let's find out.

To begin with, facts are not the same as *events* or *things*, which most people regard as the "objective," "hard-core" elements of this universe. Things and events may seem the best candidates to offer us a rock-solid foundation for our facts, but can they really fulfill their promise? True, things do exist or do not exist, and events do happen or do not happen; you can ignore things and events but not deny them. So by replacing facts with events, we might think we have found the solid objective foundation that we would like to have for our facts. However, facts and events are very different from each other. Unlike facts, events are dated, tied to space and time, whereas facts are detached from space and time. It is even considered a fact that certain events did not occur at all—for instance, it is a fact that Darwin did not have a copy of Mendel's 1866 article in his collection of papers and books. Apparently, a fact is not

---

22 For a thorough analysis see: A.R. White, *Truth* (Garden City, NY: Doubleday, 1970), chapter 6.

the same as an event; the best we can say is that a fact is a *description* of an event, but not the event itself—which makes quite a difference.

But if it is true that facts are different from events, things, situations, and processes, then this suggests that facts must be merely *thoughts*, existing only in our minds as something purely subjective. However, facts cannot be equated with thoughts either. Thoughts can have some peculiar characteristics, such as being imaginary, illogical, confused, time-consuming, and so on—whereas facts cannot. Facts on the other hand deal with what the events actually are, and not with what they might be. Facts are true, even if some people have never thought about them. Facts are always about something outside our thoughts and refer to something independent of our thinking. This makes facts unchanging entities; true, sometimes we declare something a fact, which turns out on further investigation not to be a fact, but the facts themselves cannot change. Therefore, a fact is not just a thought, but it may be the *object* of a thought.

What we just said—that facts are true, even if some people have never thought about them—may seem to contradict what we said earlier—that facts only entered the world when humans came along. But that is only seemingly so. Facts are true, always and anywhere, but they could not be expressed until humans arrived on the scene. The reason for this is that facts are non-material entities which require a human mind. Once we start thinking and talking about them, facts seem to come back into space and time. That's where we need to address a fourth element in this discussion: statements. (Of course, one could assume at any time that facts have always existed in *God*'s mind, but that is not the point of discussion here.)

Not surprisingly, some have claimed that facts are identical to what people say about them—that is, identical to (true) *statements*. However, if that were the case, there would be as many facts as there are statements (for instance, facts would be different in English and Dutch). Obviously, facts have to be clearly distinguished from statements. Statements can be hypothetical, inaccurate, exaggerated, long-winded, and difficult to understand, and so forth. Facts on the other hand cannot be any of these; a fact may be hard to accept, but never hard to understand; it is never hypothetical or half-true. There are even facts which everyone has forgotten or which were never thought of yet or which were never expressed yet in a statement. So we must come to the conclusion that a fact is not a statement, but that it may be the *content* of a statement.

From this analysis it follows that we are facing here a rather intricate situation: If facts are neither events nor thoughts nor statements, what are they? A fact is a more complex concept than it may appear at first sight. It certainly is not a "solid" entity like rocks and stones. Facts actually feature as a focus point at the intersection of those three other elements: A fact is not an event but the description of an event, not a thought but the object of a thought, and not a statement but the content of a statement. A more comprehensive definition of a fact could run like this: A fact is the description of an event, the object of a thought, and the content of a statement—and all of these at once.

Facts are closely connected to these three other elements through the process of *abstraction* and *interpretation*: Facts are interpretations of events by means of thoughts and statements. It is through interpretation that thoughts and statements transform events into facts. Facts need events so they can be tested; they need thoughts so they can be understood; and they need statements so they can be communicated. Unlike things, events, and situations with their physical nature, facts are non-material, mental entities. Since facts are not objects like stones, we cannot bump into them. Therefore, there are no "hard-core" facts, because facts cannot be touched, heard, or seen—perhaps events and things can, but facts cannot, as they are immaterial entities. Facts are *about* something "rock-solid," about something beyond our control, but they are not "rock-solid" themselves, since they have a man-made part, dependent on human interpretation. Facts are mental entities, so there is nothing physically solid about them—and yet they are not purely mental creations either, because they are about something outside themselves, which is the final touchstone of that which a fact describes (otherwise they are not facts but imaginary, fictional thoughts). It should not surprise us then that a camera, for example, cannot capture facts—all it can capture is things, situations, and events.

There is another aspect to this. Because of their man-made component, there can be various ways of looking at the same thing or event—which makes for various kinds of facts. Facts are always interpreted from a specific *perspective* and, since there are many more perspectives than what science tries to capture with its barometers, thermometers, and spectrometers, science does not give us the only window on the world; there are many other windows, views, vistas, aspects, perspectives, or whatever you wish to call them. Reality is like a jewel with many facets that can be looked at from various angles, from different viewpoints. Just as the "physical eye" sees colors in nature, so the "artistic eye" sees beauty in nature, the "rational eye" sees truths and untruths, the "moral eye" sees rights and wrongs, and the "religious eye" sees everything in relationship to God. All these "eyes" claim to be in search of reality, but each one "sees" a different aspect of it—and therefore sees different facts.

Unlike animals, which live only in a world of things, events, and situations, human beings live also in a world of facts. By stating facts, they always claim much more than what they perceive. Facts transform things of the world into objects of knowledge; they change sensations and perceptions into observations, thus enabling humans to see with their "mental eyes" what no physical eyes could ever see before. Needless to say, there would be no science without them. When humans appeared on the world scene they brought with them a rich world of facts that had remained unnoticed before.

The best we can say right now is that there is a deep rationality divide between the pre-human and human world. There is no way of telling whether this leap can be fully explained in terms of genes and mutations. Obviously, the faculty of rationality requires a body with the "right" features. Even Thomas Aquinas was aware of the fact that there is some tie between the brain and intelligence, recognizing the brain to be the organ of the imagination, the activity of which faculty is a necessary condition for thought, when he said, "it was necessary that man have the biggest brain in proportion to his body among all the animals so that the operations of the interior sense powers in him which are necessary for the operation of the intellect would be more readily exercised."[23]

Therefore, the "right" mutations were *necessary* to achieve this. But whether those mutations were also *sufficient* is impossible to tell ahead of time. Those who think the rationality divide can eventually be explained completely by science must realize this cannot be more than a conviction—more of a program than an achievement. There is no scientific way of corroborating such a conviction—it is worth as much as the opposite claim that a scientific explanation of the rationality divide will never be fully possible.

---

23 Aquinas, *Summa Theologiae*, I 91.3 ad 1.

8

The Morality Divide

HUMAN BEINGS HAVE THIS EXCEPTIONAL CHARAC-
teristic of using moral standards to judge human conduct as being morally right or wrong. Having moral judgments seems universal in humanity.
In that sense, *Homo sapiens* is also *Homo moralis*. Humans ought to
do what they ought to do. Thomas Aquinas mentions a very general
moral rule: "Good should be done and pursued, and evil avoided."[1] As
he explains, this principle serves the practical reason just as the principle
of non-contradiction serves the speculative reason. C.S. Lewis put it this
way, "human beings, all over the earth, have this curious idea that they
ought to behave in a certain way, and cannot really get rid of it."[2] They
evaluate actions as good or evil. They "intuitively" know that they owe
other human beings certain moral *duties* and that other human beings
owe them certain moral *rights*. This intuitive knowledge comes with their
faculty of morality, comparable to other faculties specific and unique to
humanity—the faculties of language and of rationality.

HOW UNIQUE IS MORALITY?

No animals seem to pass the "morality test." It looks like we have here
another deep divide between them and us. Only humans seem to have
the awareness of good behavior and bad behavior, of moral rights and
moral duties. How come?

Let me stress first that the term "morality" in this book is not another
word for social behavior. These two are very different notions. Whereas
social behavior does have evolutionary roots in the animal world, morality may not. What is it then that makes morality so different from social
behavior? Morality is not about what the world is like, but about what the
world *ought* to be like; it is not a matter of description but prescription.
It is not about how we *do* act but about how we *ought* to act. It is not
a description of what social behavior is like, but a prescription of what

1 *Summa Theologiae*, I-II, 94.2.
2 C.S. Lewis, *Mere Christianity* (London: Collins, 1997 [1952]), 21.

social behavior ought to be like. It tells us what ought to be done — by us, as a moral *duty*, and towards us, as a moral *right* — for otherwise a moral mistake would be made. Therefore, social behavior is the way it is, but it can be right or wrong in moral terms. It is essential to all of us that we can discern if human beings are doing the right thing or not. Whereas human movements are subject to physical constraints, human actions are subject to moral ones.

How could it possibly be argued that there is no morality in the animal world? Well, it is rather obvious that animals do not have a moral code that tells them what is right and what is wrong, or what they owe other animals and what other animals owe them. They have no moral code to control their drives, lusts, and emotions. They just follow whatever "pops up" in their brains — and no one has the right to morally blame them. The relationship between predator and prey, for instance, has nothing to do with morality; if predators really had a conscience guided by morality, their lives would be pretty harsh. Dogs may act as if they are "caring," but they just follow their instinct, not some moral code; dogs happen to have such an instinct, whereas cats lack it, since it is not in their genes. Dogs are known for their social behavior, but don't confuse that with moral behavior.

As a consequence, animals never do awful things out of meanness or cruelty, for the simple reason that they have no morality — and thus no cruelty or meanness. Humans, on the other hand, definitely do have the capacity of performing real atrocities. Even kids who bully others should be held accountable and disciplined, regardless of their age. In contrast, if animals seem to do awful things, it is only because we as human beings consider their actions "awful" according to our standards of morality. For example, we will never arrange court sessions for grizzly bears that maul hikers, because we know bears are not morally responsible for their actions. Since animals have no moral code, they have no duties, no responsibilities, and consequently no rights. If animals really had moral rights, other animals would have the moral duty to respect and honor those rights as well.

We, as masters of anthropomorphism, may *think* they have morality, because the doctrine of gradualism tells us we have inherited it from them. Nevertheless, animals do not have a moral code, because they are not moral beings. Professor James V. Schall, SJ makes an illuminating comparison between a man and a crocodile. When the man ruins his life by overeating, we can tell him, "Be a man!" because that includes ruling himself. But when a crocodile is satisfied because it has just consumed its tenth missionary, we don't say to the animal, "Be a crocodile!" because it *is* being a crocodile.[3] Animals do have social behavior, but they have no moral code that tells them how to behave. On the other hand, since we ourselves do have morality, we need to treat animals humanely and responsibly — not because animals have the moral right to be treated that way, but because we have the moral duty to treat them humanely. As Antoine de Saint-Exupéry wrote in *The Little Prince*, "You are responsible forever for that which you have tamed."[4]

The persistent claim of some scientists that there must be some kind of morality in the non-human, animal world — otherwise it could not show up in humanity through evolution — is in fact very problematic.[5] They base their claim on their Neo-Darwinian conviction of gradualism and then use some examples of seemingly moral behavior in the animal world to validate their claim. They remain faithful followers of Darwin's program, which says: "The following proposition seems to me in a high degree probable — namely, that any animal whatever, endowed with well-marked social instincts, would inevitably acquire a moral sense or conscience, as soon as its intellectual powers had become as well developed, or nearly as well developed, as in man."[6] This clearly promotes the evolutionary leap from social behavior to moral behavior.

To underpin their claim of gradualism, some biologists, more particularly sociobiologists, use their "magic wand" of natural selection to explain moral behavior. They like to use the example of an almost universal human taboo on incest — which says, phrased as a moral law, sexual "intercourse with very close relatives is wrong and hence forbidden." Well, these biologists would point out that inbreeding between close relatives tends to bring out recessive lethal traits and other afflictions that lessen the offspring's reproductive success. Hence, so they argue, natural selection has been

3   James V. Schall, "What Is and What Ought to Be," *New Oxford Review*, November 2018, 29.

4   Antoine de Saint Exupéry, *The Little Prince* (San Diego: Harcourt, Brace and World, 1971 [1943]), chapter XI.

5   The most extreme defender of this viewpoint is Peter Singer: Jeffrey A. Schaler, ed., *Peter Singer under Fire: The Moral Iconoclast Faces His Critics* (Chicago: Open Court Publishers, 2009).

6   Charles Darwin, *The Descent of Man*, 71–72.

promoting a genetic basis for behavioral avoidance of intercourse with close relatives; and that is where they believe this moral law stems from.

In response, one could argue that this sociobiological claim comes close to a circular argument: The cultural universality of the prohibition of incest suggests a genetic basis; since there is a genetic basis, the prohibition of incest is a universal phenomenon in human cultures. Besides, it should be pointed out that if the incest taboo were really genetic, there would be no need to consider incest immoral, let alone declare it illegitimate through a moral code, for genetics took care of that already. Yet incest does happen among people in spite of the genes that are supposed to prevent it.

What sociobiologists are good at is re-defining terms. In this case they equate the moral term "incest" with the biological term "inbreeding." However, more detailed moral rules as to what is considered incest clearly vary between different cultures. In Ptolemaic Egypt and Incan Peru, for instance, incest was as much as the "rule" in the royal line. Some cultures prohibit sexual relations between clan members, even when no traceable biological relationship exists. In many cultures, certain types of cousin relations are preferred as sexual and marital partners, whereas in others these are considered taboo. Some cultures permit sexual and marital relations between aunts/uncles and nephews/nieces. It is only parent-child and sibling-sibling unions that are almost universally taboo.

Another example some sociobiologists use to make their case are the moral laws given in the fifth and sixth commandments of the Decalogue, "You shall not kill" and "You shall not commit adultery." Some biologists have argued that humans have become monogamous "by nature," since that would give the offspring a better protection, and that is why natural selection must have promoted monogamy—thus giving the sixth commandment a biological basis. Others have made the claim that killing any members of the same species would undermine the persistence of the species—hence the prohibition was selected for by the process of natural selection. They even consider the moral value of paternal care for children to be a product of natural selection, since fathers who do not feel an "instinctive" responsibility towards their underage children would reduce their offspring's reproductive success. What all these cases have in common is that morality is supposedly based on genes promoted by natural selection.

However, these claims are hard to make. Moral laws such as "You shall not kill" or "You shall not commit adultery" could hardly make it through natural selection, for their offenders — the killers, the promiscuous, and the rapists — would do much better in reproduction than their victims or than the ones who go by a moral code. Moral laws do not seem to have much survival value and therefore are not good candidates for natural selection. Some moral issues such as altruism are actually foes of natural selection. Altruism is no good friend to survival; it actually amounts very often to "genetic suicide." Whereas natural selection is based on self-preservation at the cost of others, morality is often self-sacrifice for the good of others. When firefighters, or soldiers on the battle field, die in the line of duty, they typically do not give their lives for the sake of natural selection. Francis Collins, the former Head of the *Human Genome Project* and currently Director of the *National Institutes of Health*, made very clear that morality goes against natural selection: "Evolution would tell me exactly the opposite: preserve your DNA. Who cares about the guy who's drowning?"[7]

Although not a moral duty *per se*, altruism is in essence moral behavior for the sake of serving others, without expecting any advantage. This is what we really and rightly call "unselfish altruism." The notion of charity, for instance, is completely about giving for giving's sake. The same is true of donating blood to strangers: It does not help relatives, not even oneself, so it is not subject to natural selection. Moral issues like these are in a constant conflict with our urge to survive, to care for ourselves, and to provide for our offspring — which shows us again that morality is no good friend of survival. What we actually *do* achieve, which may be a biological target for natural selection, may be very different from what we *want* to achieve, let alone what we *ought* to achieve. What one ought to achieve (which is a moral value or rule) is not necessarily what one wants to achieve (which is a motive or intention), and what one wants to achieve is not always what one actually does achieve (which is a mere effect).

C.S. Lewis makes some interesting remarks in this context. Just as he rejected a Darwinian explanation for the human mind because it undermines the validity of rationality itself, so did he reject a Darwinian

---

7 "The Question of God — an interview with Francis Collins," *PBS*, 2004.

account of morality because it would undermine the authority of morality by attributing it to an essentially amoral process of survival of the fittest. In his book *Miracles* he made this point even more explicit: A Darwinian process "may (or may not) explain why men do in fact make moral judgments. It does not explain how they could be right in making them. It excludes, indeed, the very possibility of their being right."[8] After all, if moral judgments are ultimately the products of natural selection, then all such judgments must be equally valid or invalid—that is, equally right or equally wrong. In the world of natural selection, there is no right or wrong—only different degrees of survival and success.

Ironically, sociobiologists try to undermine morality by pointing out that altruism—giving up personal comfort for the *benefit* of others—can be found even in the animal world where natural selection reigns—that is, the principle of increasing one's own reproductive success at the *expense* of others. Again, sociobiologists use their tactic of re-defining the moral term "altruism" with a biological version of it. Their favorite example is a colony of bees in which sterile worker bees "unselfishly" help the queen raise her own progeny. Sociobiologists would explain this kind of "altruism" as a form of helping one's close relatives, who carry DNA very similar to one's own. Since natural selection is a matter of balancing the benefits against the costs, they explain even altruism as a way of promoting one's "own" DNA by diminishing one's own offspring (a cost) but increasing the offspring of one's relatives (a benefit)—which they dub "kin selection."[9] But it is very confusing, to say the least, to equate biological altruism, which has the *effect* of benefitting others, to moral altruism, which has the *intention* of benefitting others—which makes quite a difference.

It is this mechanism of kin-selection that enticed the British biologist Richard Dawkins to write his book *The Selfish Gene*. In this book he declares on the first page, "My purpose is to examine the biology of selfishness and altruism." Since it was natural selection that promoted "altruism" in the world of animals by a purely "selfish" mechanism, it must have done the same, so his argument goes, in the world of humans. Really? Well, first of all, this phenomenon may be called "altruism"—with

8   C.S. Lewis, *Miracles* (London: Fontana, 1963), 50.
9   J. Maynard Smith, "Group Selection and Kin Selection," *Nature*, 1964, 201 (4924): 1145–47.

a deflated kind of terminology—but this is not altruism as it is understood in morality. Second, speaking of "selfish genes" is as silly as speaking of "selfish atoms." It is merely a rhetorical trick used by Dawkins to make his case. Third, Dawkins shrewdly manipulates the terms he uses for his explanations. He borrows the term "selfishness"—a term that makes perfect sense in a moral context—then applies it to genes in genetics, and finally uses the "selfishness of genes" to eliminate altruism in a moral context as being a mere battle of selfish genes. That looks like he got caught in what we call a circular paradox.

In contrast, the biological metaphor of "selfishness" is only possible because we already have an understanding of human behavior in a *moral* context. Dawkins turns things around: He treats the biological metaphor of selfishness as the real thing, and then uses it again to discard the actual real thing—the human behavior of altruism—as being unreal. He makes the gene the basic unit of "selfishness." However, the language of selfishness does not make sense in genetics, because genes cannot be selfish or unselfish the way human beings can be. Yet he uses the expression "selfishness of genes" to get rid of altruism in the world of human beings. The assessment of the anthropologist Donald Symons might be right on target: "[T]he rhetoric of *The Selfish Gene* exactly reverses the real situation: through [the use of] metaphor genes are endowed with properties only sentient beings can possess, such as selfishness, while sentient beings are stripped of these properties and called machines.... The anthropomorphism of genes ... obscures the deepest mystery in the life sciences: the origin and nature of mind."[10] We are back to where we started.

A MORAL CODE OR A GENETIC CODE?

In morality, we speak of "moral duties" and "moral rights" which are based on a moral code. Could this moral code be etched in our DNA code? Could we be "hard-wired" to be moral and make specific moral judgments to determine that some acts are good and others bad? It is very tempting to think so, especially from a gradualist viewpoint.

We seem to be dealing here with an *empirical* issue. If so, only time can tell whether there are certain genes that make us moral and determine

10 Donald Symons, "The semantics of ultimate causation," in *The Evolution of Human Sexuality* (Oxford: Oxford University Press, 1981), 42.

which moral code we use. Right now, no such genes have yet been found. However, we may not have to wait for new genetic discoveries, as we might be able to decide this issue ahead of time—*a priori*, that is.

An *a priori* approach like this would not be as uncommon as it might look. Sometimes the tools of logic and philosophy can tell us ahead of time—*a priori*, that is—what science can and what it cannot achieve. Let me explain this with the following case that I borrowed from the late Nobel Laureate and physiologist Peter Medawar.[11]

Anyone looking at paintings of the famous Renaissance painter El Greco will notice that most of his figures are unnaturally tall and thin. Some scientists were eager to explain this in scientific terms. One of them came up with the hypothesis that El Greco must have had a form of astigmatism that distorted his vision and led to elongated images forming on his retina. Sounds interesting, but this hypothesis is doomed from the very beginning. Even if El Greco did see the world through a distorting lens, the same distortion would apply to what he saw on his canvas. These two distortions would cancel each other out, and the proportions in pictures would remain realistic. So we must come to the conclusion that El Greco's figures, particularly the holy ones, appear unnaturally thin and tall because that was his intention, not forced by any gene or disorder. In other words, he painted them that way on *purpose*.

Imagine how these scientists had spent much time, energy, and funding on testing their hypothesis in the lab, not realizing that it was bound to fail ahead of time on purely philosophical, or at least logical, grounds. Yet, some did not give up that easily and tried to also attribute Vincent Van Gogh's preference for yellow colors to a visual disorder called xanthopsia; others mentioned drug use or glaucoma as the causative agent. Even if Van Gogh did view the world through a yellow filter, he would also view the colors on his canvas that way. Instead, he must have just chosen the yellow color for some reason—on purpose, that is!

In both cases, science just fails us, since philosophy and logic are able to show us *a priori* that the scientific explanations mentioned above are doomed to fail, even before any scientific test has been done. There are certain "scientific" ideas one should not spend time, money, and energy on, because philosophy and logic can prove they are headed for a dead

11 Peter Medawar, *Advice to a Young Scientist* (New York: Basic Books, 1981), 9.

end. They are inventions that could not possibly make it to discoveries, purely on logical or philosophical grounds.

Something similar might be the case when it comes to the genetics of morality. All scientific endeavors to reduce a moral code to a genetic code might be doomed to fail ahead of time merely for philosophical or logical reasons. Sometimes we know ahead of time — without doing any genetic research — that, in a case like this, a genetic explanation does not make sense or is not even possible. How do we know so?

Let's start with the idea of what they call "genetic determinism." People who believe in strict genetic determinism — and a number of geneticists do, albeit a minority probably — are of the opinion that morality must, like anything else, be determined by genes and DNA. Here are two of their powerful advocates. Sidney Brenner, one of the DNA pioneers, said not too long ago he could compute an entire organism if he were given its DNA sequence and a large enough computer.[12] With a like mind, the American molecular biologist Walter Gilbert had the audacity to claim that "when we have the complete sequence of the human genome we will know what it is to be human."[13] That is genetic determinism in full glory!

What is wrong with these statements? Not only is genetic determinism a rather unscientific theory — it steps outside science to claim that there is nothing outside the realm of science, which is a claim that could never be experimentally or scientifically tested (see chapter 12) — but also it is a rather questionable claim from a philosophical and logical point of view. One of its problems is that it leads us into logical trouble. This is one of those cases where we know ahead of time that exhaustive genetic explanations do not make sense or are not even possible. Let's see why not.

If we have a free will — and morality assumes we do, so we can make moral decisions and act accordingly — you might wonder whether this free will could perhaps be genetically determined, based on a gene or a series of genes. Theoretically there could indeed be genes that determine what we choose and what our moral code is, but if these genes would also determine the outcome of these choices, then we cannot really make

12   He made this claim at a symposium in commemoration of the 100th anniversary of the death of Charles Darwin.

13   Walter Gilbert, "A Vision of the Grail," in *The Code of Codes: Scientific and Social Issues in the Human Genome Project*, eds. D.J. Kelves and L. Hood (Cambridge, MA: Harvard University Press, 1991).

free choices and have lost the free will we thought we had. Is it possible to eliminate our free will through genetics? Philosophy and logic may help us answer this question ahead of time, without our having to do any research in genetics. They tell us that a radical form of genetic determinism would cause philosophical trouble. There are at least three arguments that plead against universal determinism.

The first argument is that a universal form of determinism would lead us into a vicious circle of infinite regress—with no way to get in or out. Here is how. Determinism posits an infinite regress of causes—cause n is caused by cause $n-1$, and so on backwards in time or space. Say, quantum particles are predetermined, but then the question arises "determined by what?" By smaller particles? What then determines these particles? Even smaller particles? The chain of causes would have to regress infinitely—which is in itself problematic because an infinite number of things cannot actually exist in a material world.

A second argument goes as follows. Universal determinism creates a loop that makes for a logical paradox caused by the problem of self-reference. Self-reference is used to denote a statement that refers to itself. The most famous example of a self-referential sentence is the so-called liar sentence: "This sentence is not true." If we assume the sentence to be true, then what it states must be false. If, on the other hand, we assume it to be false, then what it states is actually true. In either case we are led to a contradiction. Well, trying to convince someone of the truth of universal determinism smacks also of self-refutation. How could this claim change someone's mind if everything is fully predetermined anyway? Ironically enough, people who defend universal determinism—let us call them determinists—are willing to spend their entire career on forcing anyone else to choose their deterministic conviction that human beings cannot choose. Chesterton once said in a more direct and simple way that, if the world is determined, it makes no sense to say "thank you" to the waiter for bringing the mustard. To give thanks implies that something that did happen need not have happened.

The third—and arguably the strongest—argument against determinism is that a world-view of universal determinism makes for a boomerang effect. If genes, for instance, really determine everything in one's life, then they would also determine one's decision to believe or not to believe the claim that genes determine everything in life. The key problem is that we are dealing here with *beliefs*, and beliefs are not material entities like genes—for unlike genes, they can be true or false. If I believe that genes determine everything, I have no reason to suppose my belief is true, and hence I have no reason for supposing that genes determine everything. This is a "boomerang theory" in *optima forma*—that is, it defeats itself, for once we consider it to be true, it becomes false.

Let's shift the focus from determinism back to morality with the following question: What determines morality—is it a genetic code or is it a moral code? Gradualists would opt for a *genetic* code that we ultimately inherited from the animal world, whereas moralists plead for a *moral* code that only humans have access to. A genetic code would make us act a certain way "by nature," whereas a moral code would make us do "by choice" what our genes do not make us do "by nature." How do we decide which option is right? We will argue here, with several reasons, for the primacy of a moral code over a genetic code.

Reason #1: It is hard to believe that all those people who are willing to break a moral rule when they can get away with it are acting against their genes. There are too many people who ignore what their genetic code is supposedly telling them. Bad moral behavior can spread like wildfire, but mutated genes don't spread that quickly. As a matter of fact, there are too many parents who ignore what some claim is an "inborn" or "genetic" responsibility of parenting. There are too many spouses who violate the sixth commandment, "You shall not commit adultery." Too many folks also violate the fifth commandment, "You shall not kill." When it comes to moral laws, everyone knows about them and yet everyone breaks them again and again. Genes do not seem to prevent this.

Those who go against moral laws or ignore them may be steered by drives and passions, but it is very unlikely they are controlled by their genes. Instead, the opposite could be argued: Morality has the power to overrule what our genes dictate—passions, emotions, and drives. Perhaps genes contribute to our being moral beings, but they do not and cannot dictate specific moral laws, rules, and values. To use an analogy, genes may help create good or bad football players, but that is not how the rules of the game are regulated. What is right or wrong in a moral sense is not determined by genes, for genes are material entities that cannot possibly make anything good or bad in a moral sense. There is no good or bad, right or wrong in the material world. Besides, when people change their moral code, would that mean their genes have changed? Genes may make us act a certain way, but whether such an act is morally right or wrong is a completely different issue — a moral issue, not a genetic one.

Reason #2: Those who reduce a moral code to a genetic code should ask themselves why we need a moral code to do what we would do or not do "by nature" anyway.[14] A morality that is supposedly preprogrammed in our genes would make a moral code completely redundant. If morality can really be reduced to what we do "by nature," there would obviously be no need for a moral code as well. We would all act rightly merely by nature, so it would not even be possible to do something morally wrong. This makes us realize there must be a moral code beyond and above a genetic code. We do need a moral code because God, according to St. Augustine, "wrote on the tables of law what men did not read in their hearts."[15] Since we are not moral by nature, morality has to be taught and nurtured, above all by the Scriptures and the teaching authority of the Church.

In contrast to a biological explanation of morality, one could very well argue that moral laws tell us to do what natural selection does *not* promote and what our genes do *not* make us do "by nature." If moral behavior were genetic, we would all act "rightly" by mere nature, so it would not even be possible to do something morally wrong. Of course, one could counter that some of us might have mutated genes that may direct us to do what is wrong. But if that were the case we would have no reason anymore to speak of right or wrong, for either way would be a preprogrammed outcome that would release us from any moral responsibilities.

Reason #3: Reducing morality to a product of natural selection must face the problem that morality is not survivor-friendly. Most moral laws do not seem to have any survival value and therefore cannot be the target of natural selection. As we said earlier, the offenders of moral laws — the killers, the liars, the rapists, and the promiscuous — reproduce much better than their victims. Apparently, morality and "survival of the fittest" do not go well together. Natural selection is about success at the *expense* of others; morality is about duties to the *benefit* of others. Natural selection *eliminates* the ones who cannot care for themselves; morality instead *cares* for those who cannot care for themselves.

14 Cf. R. C. Lewontin, "The Fallacy of Biological Determinism," *The Sciences*, 1976: 6–10.

15 St. Augustine, *En. in Ps.* 57, 1: PL 36, 673.

Whereas it is hard to disregard what a genetic code enforces, a moral code can be ignored at any time. When it comes to moral laws, everyone knows about them and yet everyone repeatedly breaks them. Unlike the laws of nature, moral laws can in fact be ignored (try to do that with the law of gravity!). Mothers who abandon their babies are very unusual in the animal world, because of genetic constraints, but in human societies they are not quite unusual, because maternal responsibility is a moral law that can easily be ignored or sidestepped. Natural selection on the other hand would go against abandonment of babies by their mothers. Since we do not have a moral code determined by a genetic code, we have the freedom to accept or reject what is morally right and what is morally wrong — as we see happening all the time — because right and wrong are not under the control of genes. Morality gives us the power to overrule any "innate" drives and emotions if we deem them immoral, but that's a choice we ourselves have to make; it's not enforced on us by genes.

Reason #4: To claim that morality is "nothing but" a product of natural selection is actually a self-defeating activity. It is basically suicidal: the snake of this theory is swallowing its own tail, or rather its own head. If indeed we were to claim that morality is nothing but a "pack of genetic instructions," this very claim would not be worth more than its molecular origin, and neither would we ourselves who are making such a statement. The very claim that everything is a matter of genes would make it a purely genetic issue, so it could no longer be possible even to meaningfully make a claim that moral claims are genetic by nature. That would be the end of any truth claims, even in morality.

Claims of "nothing-buttery" in matters of rationality and morality just defeat and destroy themselves. They undermine their own truth claims by cutting off the very branch that the person who makes such claims is — or actually was — sitting on. This conclusion should put a science such as sociobiology in its proper place: it may be a fantastic specialty in the field of science, but there must be more to life than genetic instructions in charge of human behavior — unless the enterprise of sociobiology itself is entirely a product of genetic instructions too. We are certainly not required to take the "survival of the fittest" law as a moral guideline — we are actually not allowed to on purely moral grounds.

Reason #5: Believing that morality is in the genes is a very shaky belief that undermines its own foundation and creates a paradox. Some people, such as the biologist J.B.S. Haldane and the philosopher C.S. Lewis, have worded this paradox as follows: "If my mental processes are determined wholly by the motions of atoms in my brain, I have no reason to suppose that my beliefs are true...and hence I have no reason for supposing my brain to be composed of atoms."[16] Besides, as Francis J. Beckwith warns us, we have to face the following problem: "[I]f your belief in the moral law can be attributed entirely to our genes tricking us into believing that there really is a moral law, why not extend that same analysis to all other beliefs that arise from our mind?"[17] So we would end up with many more illusions fobbed on us by our genes—science and math, to name just a few. Do we really want to pay that price? That would amount to "losing your mind."

If we are looking for a key to understanding humans and their moral beliefs, this key will not be found in something material, such as genes, but in something immaterial, the mind. Morality comes from the immaterial mind, not from the material brain or genome. The brain is governed by laws of physics, chemistry, and biology, but thoughts and beliefs are not. It should not surprise us then that people have known the contents of their own minds from time immemorial without knowing anything about brains and genes. They knew also about morality without knowing anything about brains and genes. Claiming differently reduces the working of the mind to the materialism of the brain. That is basically what Charles Darwin did when he said in an early private notebook, "Why is thought, being a secretion of brain, more wonderful than gravity as a property of matter?"[18] This boils down to attacking the power of reasoning and logic by using the power of reasoning and logic—a textbook case of circular reasoning.

Reason #6: When reducing morality to genetics, we cannot have it both ways. First, evolutionary theory tells us that our moral behavior is inborn and that its reproductive success is based on our *believing* that morality is objective. And next it tells us that morality is not objective

16 J.B.S. Haldane, *Possible Worlds and Other Essays* (New York: Harper and Brothers, 1928), 209. Also in *The Inequality of Man* (London: Chatto and Windus, 1932).

17 Francis J. Beckwith, "On God, the Moral Law, and Losing Your Mind," *The Catholic Thing*, December 9, 2016.

18 Paul H. Barrett, et al. (eds.), *Charles Darwin's Notebooks, 1836–1844* (Cambridge: Cambridge University Press, 1987), 291.

at all, in spite of the fact that all of us supposedly have an inborn belief that morality is objective. If we were really able to uncover the illusion of morality, morality would lose its evolutionary power immediately.

Thus we end up with another contradiction. The evolutionary theory's success depends on our believing that morality is objective. It is because we desire to act in accord with this belief that we presumably forego the pursuit of our own interests for the good of others — even when we can escape detection and punishment. If this theory is true, then the assumed objectivity of morality could only play its evolutionary role if we remained ignorant of the theory. Even if we happen to come in contact with the theory, we would still find ourselves pushed by a belief that contradicts it. As a matter of fact, the evolutionary approach is not an explanation of morality; it's a denial of morality. It explains why we think moral truths exist when, in fact, they don't. That would be a troublesome outcome.

Reason #7: What biologists, especially the sociobiologists among them, basically do is reduce morality to a mere issue of matter, more specifically, of genes and their DNA molecules. This has materialism written all over it. Materialism claims the physical world is all there is. It emphatically proclaims that everything that exists is matter, and that matter is all there is. But what entitles us to believe that everything is matter and that everything is material? The claim of materialism would at best be a dogmatic conviction, certainly not backed by science itself. Science on its own can never prove that matter is all there is, because it first limits itself to matter and then says there is nothing but matter. If materialism is true, we cannot even know that it is true (see chapter 11).

Matter may be everywhere, but it is certainly not all there is. If matter were indeed all there is, then one should wonder what materialism itself is. Another piece of matter? If not, then there must be more than matter. This definitely leaves room for non-material things such as logic, mathematics, philosophy, and ultimately religious faith. So why then not for morality as well? Morality is about what is good or bad, right or wrong. In the world of matter, things are large or small, light or heavy, hard or soft, but never good or bad, right or wrong. There are no "oughts" in the material world. Apparently there is so much in life that the thermometers and Geiger-tellers of materialism can never capture — things such as oughts, thoughts, values, beliefs, laws, experiences, hopes, dreams, and ideals. There is no way materialism can deal with these—other than denying them, but then it must deny itself as well.

Reason #8. In fact, morality is about something that is outside the scope of biology, actually beyond the reach of the natural sciences. Biology is blind to moral values, so it cannot possibly discern anything that is on its "blind spot." Therefore, science cannot monitor morality, but it is rather the other way around—morality ought to control science instead. Nazi-doctors such as Joseph Mengele show us what happens when morality does not control their scientific research. Albert Einstein was right when he spoke of "the moral foundations of science, but you cannot turn around and speak of the scientific foundations of morality.... [E]very attempt to reduce ethics to scientific formulae must fail."[19] Morality can interrogate science, but science cannot question morality—it lies beyond its reach.

When we make "survival of the fittest" the driving force of morality, we end up with what became known as eugenics—a brutal movement that inflicted massive human rights violations on millions of people. The "interventions" advocated and practiced by eugenicists involved a wide range of "degenerates" or "un-fits"—the poor, the blind, the mentally ill, entire "racial" groups such as Jews, Blacks, Roma ("Gypsies"); all of these were deemed "unfit" to live according to the despotic dogma called "survival of the fittest." And this in turn led to practices such as segregation, sterilization, genocide, preemptive abortions, euthanasia, designer babies, and in the extreme case of Nazi Germany, mass extermination. That's what happens when biology controls morality, instead of the other way around.

Reason #9: How could genes ever make something obligatory? There are no "oughts" or "rights" or "duties" in the material world of genes. A gene cannot make something obligatory—perhaps more or less effective, more or less successful, or whatever, but never more or less obligatory, let alone morally right or wrong. What is right or wrong in a moral sense is not determined by genes, as genes cannot make anything good or bad, right or wrong. Genes may make us act a certain way, but whether such an act is morally right or wrong is a different issue—a moral issue, not a genetic one. And from this follows something else: Because there are

19   In a discussion on science and religion in Berlin in 1930.

no "oughts" in the world of genes, there are no "oughts" in the world of non-human animals.

As Michael A. Simon remarks, "In order for a human trait to be explained biologically, it must first be 'biologized'. . . . The problem with such biological reduction is that it is likely to sacrifice precisely those features of human social behavior that give it a socially or philosophically distinctive character."[20] So by "biologizing" morality, we inevitably lose its distinctive moral character. To get its distinctive moral character back, we must leave the territory of genetics and enter the domain of human morality.

Let's come to a conclusion: If the above reasons do not convince you on a case by case basis, perhaps combined they make for a strong plea against the claim that a moral code is ultimately nothing else but a genetic code. The best we can say right now is that there is a deep morality divide between the pre-human and human world. There is no way of telling whether this leap can be fully explained in terms of genes and mutations. Obviously, the faculty of morality requires a body with the "right" features. The "right" mutations were probably *necessary* to achieve this, but as to whether those mutations were also *sufficient* is impossible to tell ahead of time. Those who think the morality divide can eventually be explained completely by science must realize that this cannot be more than a conviction — more of a program than an achievement. There is no scientific way of corroborating such a conviction — it is worth as much as the opposite claim that an all-inclusive scientific explanation of morality will never be fully possible.

WHERE DOES THE MORAL CODE COME FROM?

Humans ought to do what they ought to do. Yes, but what exactly it is that they ought to do is a different issue — the issue of what in particular it is that we owe others and others owe us. As was said before, what we owe others, in moral terms, are our *duties*; what others owe us are our *rights*. Duties and rights have a naturally reciprocal relationship — they go hand in hand and keep each other in tow. The right to defend one's own life comes with the duty to protect someone else's life; the right to

---

20 Michael A. Simon, *The Matter of Life: Philosophical Problems of Biology* (New Haven, CT: Yale University Press, 1971), 225.

seek the truth goes with the duty to let others seek the truth; the duty to honor the dignity of human beings is what we owe others as a duty, and it is something others owe us as a right. In other words, no duties no rights, and no rights no duties. No one has the duty to have children, so no one has the right to demand children. "Duties" and "rights" are very specific moral concepts.

Morality does not come with a special race, nation, party, or church — it is a common property that belongs to all of us who belong to the species *Homo sapiens*. However, although the faculty of morality may be universal among humans, that does not mean moral duties and moral rights are identical across all of humanity. There seem to be different moral codes in different parts of the world — at least to a certain extent. Indeed, there are some important disagreements about what exactly is morally good or right when we look at different cultures. Nonetheless, beneath all disagreements about lesser moral laws and values, there lies always an agreement about more basic ones.

To use the analogy of different languages: Beneath the different words of different languages we find common concepts — and this is what makes translation from one language to another possible. In a similar way, beneath different social laws, we find common moral laws. As a matter of fact there is not a great deal of difference between Christian morality, Jewish morality, Hindu morality, Muslim morality, Buddhist morality — although there's a great difference between these religions. Perhaps the best-known universal moral principle is the so-called Golden Rule.[21] In its negative form it says "Do not do unto others as you would not have them do unto you." Its positive form is "Do unto others as you would have them do to you." This Golden Rule can be found in Christian, Jewish, Islamic, Buddhist, and Confucian texts, among others. Perhaps the oldest moral code is the Code of Hammurabi from around BC 2250. C. S. Lewis published a list of universal moral principles which he called "Illustrations of the Tao or Natural Law."[22]

It is mostly through Thomas Aquinas that this concept of moral communality has become known as the *natural law*. Its key idea is that moral laws are based on human nature, on the way we *are*. As a consequence,

---

21   Also found in the Bible: Tobit 4:15; Matthew 7:12; Luke 6:31.
22   In the appendix of his book *The Abolition of Man* (1944).

morality is a function of human nature, so that reason can discover valid moral principles by looking at the nature of humanity and society, no matter where on Earth. This means that what we ought to *do* is related to what we *are*. "You shall not kill," for instance, is based on the real value of human life and the need to preserve it. "You shall not commit adultery" is based on the real value of marriage and family, the value of mutual self-giving love, and children's need for trust and stability. We share these moral convictions to some degree with all of humanity. Every culture in history has had some version of the Ten Commandments.

Natural Law rests upon the claim that things have natures and essences which all of us can detect and to which our actions can correspond. All things possess a nature or essence; they flourish when they are treated in accord with that nature or essence—and they wither when they are not. There are many reasons for making this claim. One is the fact that all things act in a predictable fashion; for example, when we learn the properties of oil and water, we can predict certain things about their behavior. In a similar way, Natural Law holds that we live in a Universe of things that have a nature to them and that we shall get the best out of these things if we act in accord with the nature that is written into them. Abraham Lincoln spoke of "an abstract truth applicable to all men at all times."[23] Thomas Aquinas puts it this way: "the light of reason is placed by nature in every man to guide him in his acts."[24] Although the Natural Law is not universally obeyed, or even universally acknowledged, it is still universally binding and authoritative.

When we act in a *rational* way, we act in accord with our own nature and reality and in accord with the nature and reality of other things. This holds also when we act in a *moral* way. That's where the human facilities of rationality and morality meet. Even Aristotle knew that we are creatures able to give reasons regarding matters of right and wrong. Moral laws have a certain foundation in reality—they are based on relationships between human beings and things. Where, for instance, does the moral law of loving one's parents come from? The answer is that the physical and mental constitution of human beings happens to be such that children ought to love their parents in order for them to prosper as

---

23   Letter to Henry L. Pierce and others (Springfield, IL, April 6, 1859).
24   Aquinas, *De Regno ad Regem Cypri*, I, 1, 4.

human beings. Were our human constitution differently structured, we most likely would have different morals.

This idea is taught not only by the Catholic Church but in essentials by all the world's major religions and nearly all pre-modern philosophies. It is the idea that the laws of morality are not rules that we invent but principles that we discover, similar in a way to the laws of a science such as physiology. Just as our physiological nature makes it necessary for us to eat certain foods and to breathe oxygen for our bodies to be healthy, so our moral nature makes certain moral rules and laws necessary for our souls to be healthy. In other words, not only is there a physical order in nature, there is also a moral order. As Chesterton observed, "You may, if you like, free a tiger from his bars; but do not free him from his stripes. Do not free a camel of the burden of his hump: you may be freeing him from being a camel."[25]

Seen in this light, there seem to be many parallels between morality and rationality, between the Natural Law and the laws of nature.

1. It is plainly *evident* that there is a natural order in this world and that like causes produce like effects—there is just no hard proof for it. You cannot prove order in nature, you must assume it. It is equally *evident* that there is a moral order in this world. For example, it is evidently wrong to kill another human being—but there is no hard proof for it in the way science requires.

2. Rationality is in search of *universal* laws and *objective* truths in this Universe, as these tell us the way it is in this world—no matter what, whether we like it or not, whether we feel it or not. Truths are true even when we do not know yet they are true. Truths are not dependent on our knowledge and are not created by our knowledge. In a similar way morality is in search of universal rights and duties in this world; these tell us what we ought to do—no matter what, whether we like it or not, whether we feel it or not. Rights are right, even when we do not know yet they are right or do not like it that they are right.

3. Laws of nature and moral laws are both *absolute*. Scientists, for instance, are in essence absolutists: They are ultimately in search of absolute, objective, universal laws of nature, but they also know they may not have reached that point yet—that's why they keep searching, verifying,

---

25   G. K. Chesterton, *Orthodoxy*, chapter 3, "The Suicide of Thought."

and falsifying. In morality, we should strive for something similar. Just as we may be unaware of certain laws of nature that we do not know yet, so we may violate moral laws we are not yet aware of. Relativists in morality, on the other hand, defy themselves when they make the absolute statement that everything is relative. Deciding on what is true and on what is right is ultimately not a matter of opinion.

4. We cannot even establish rationality by showing how irrational it would be to reject rationality, since that would presume already some sense of rationality. Neither can we establish morality by pointing out how immoral it would be to reject morality, for that would already require a basic sense of morality. As human beings we have a sense of rationality and a sense of morality—inborn if you will, but not necessarily genetic.

5. Our genes do not determine how we choose in rationality between true and false, or in morality between right and wrong. There is no true and false, no right and wrong in the world of genes. It seems to be quite the opposite: Rationality and morality have the power to overrule what our genes dictate. In fact, they both persistently attempt to distance us from what we are or would be if our genes were in full control.

6. If you have ever been on jury duty, you know how much jurors depend on their human faculties of rationality and morality. Unlike animals, humans are able to make rational and moral decisions—which fact also entails the possibility of irrational and immoral decisions, of course. Take rationality and morality away from us, and we are merely non-human animals. Only human beings can curb their animal drives and instincts with rationality and morality.

In short, there are some strong resemblances between moral laws and laws of nature. Both are universal (applicable to everyone everywhere), absolute (without any exceptions), eternal (even if we do not know the law behind it yet), and objective (a given, independent of us and of any human authority). They are objective, universal, eternal, and absolute standards—no matter whether we are talking rationality, in terms of true and false, or morality, in terms of right and wrong.

Clearly, this conclusion may be hard to accept for moral relativists. They would object to the idea of an absolute, timeless morality by claiming that moral rules and values have repeatedly been subject to change during the course of human history. However, there is a mix-up here between moral *values* and moral *evaluations*. Moral evaluations are our personal feelings or discernments regarding moral values and laws at a certain point in time. Moral relativists think that, in making moral evaluations, we create moral values in accordance with these evaluations. So when evaluations change, the moral values and laws are said to change as well. Could that be true?

In response to this position of moral relativists, it should be emphasized that evaluations are merely a reflection of the way we discern absolute moral laws and values at a specific place and time. Whereas moral evaluations may be volatile and fluctuating, moral values and laws are timeless and unchangeable. That is the reason why we can disagree about certain moral evaluations, assuming some are true and others false. Think of the following comparison: our current understanding of physical or biological laws constantly needs revision each time we reach a better understanding of those laws. Yet, in the meantime, we assume there are timeless laws of nature, although we may not yet have fully captured them in our current understanding and in our contemporary evaluations. Something similar holds for moral laws. Besides, we may know what is morally right or wrong, but we may not be *willing* to do what is morally right.

We could illustrate this point a little further. A few centuries ago slavery was not evaluated as morally wrong, but nowadays it is by most people. Had the slaveholders won the Civil War, we might see it today as an admirable institution according to moral relativists. Did our moral values change? Our evaluations certainly did, but that does not mean moral values did too. Only some people in the past—heroes such as St. Cyprian, St. Gregory of Nyssa, St. John Chrysostom, St. Patrick, St. Anselm, St. Vincent de Paul, to name just a few—were able and willing to discern the objective, intrinsic, and universal value of personal freedom and human rights (as opposed to slavery), whereas many of their contemporaries were blind to this value (or they were not willing to do what is morally right). That is why Martin Luther King Jr. could say, "A just law is a man-made code that squares with the moral law, or the law of God. An unjust law is a code that is out of harmony with the moral law. To put it in the terms of Thomas Aquinas, an unjust law is a human law that is not rooted in eternal and natural law."[26]

---

26 "Letter from Birmingham Jail," April 16, 1963.

After all we have seen so far, it is hard to believe that the moral code is merely a matter of a genetic code. Access to a moral code, to a world of "oughts" and "rights" and "duties," seems to be a unique faculty of human beings, setting them apart again from the pre-human animal world. There is a divide here that is nearly impossible to remove. Might there be any genes involved? Only time can tell, but even if we do find such genes, we can almost predict ahead of time on logical and philosophical grounds that they would play only a very minor role.

9

The Self-Awareness Divide

NOT SURPRISINGLY, GRADUALISTS ASSUME THAT self-awareness in humans must have roots in the pre-human world too, especially in our closest relatives. They remain faithful to their founder Charles Darwin, who could never resist the temptation of gradualism—not even when it comes to self-awareness. Darwin himself had shown mirrors to orangutans to test their self-awareness, but they didn't figure the mirror out, at least not while he was watching. In 1889, German researcher Wilhelm Preyer became the first to posit a connection between mirror self-recognition and "self-awareness" in humans, and possibly in animals.

WHAT IS SELF-AWARENESS?

Self-awareness is the ability to recognize oneself as an individual separate from one's surroundings and from other individuals. Self-awareness implies that we know we remain the same person, even though our cells are constantly being replaced. Losing or replacing larger parts of my body may happen without losing my "self." I can get a new heart, liver, lungs; I can also get knee, hip, and ankle replacements; I can receive prostheses for hands and feet, arms and legs, and so on. Yet I know I am still my own old "self." I also know, for instance, that I myself started at one point in life as a fertilized egg cell and that at some point in time I will be dead. It is my very "self" that connects all the stages of my life as one long continuum. The Anglican theologian and physicist John Polkinghorne is right: "The atoms in each of us are being continually changed by eating and drinking, wear and tear. They cannot be the source of our experience of a continuing self."[1] In other words, although my body changes constantly, I myself do not—that is, my identity remains the same. A fancier word for all of this is self-awareness.

After what we have seen in the previous chapter it is obvious that the issue of altruism and selfishness is closely connected to the issue

---

1   John Polkinghorne, *Science and Theology* (Minneapolis, MN: Fortress Press, 1998), 62.

of self-awareness. Without self-awareness there could not really be self-ishness and altruism, for "self" separates me from others. So what is it that we know about self-awareness? What is this "self" that we are supposedly aware of?

The idea of "I" and "self" has been under attack for quite a while. More recently it was the late British philosopher Gilbert Ryle who spoke of the "systematic elusiveness" of "I."[2] Ryle tried to paint the "self" as something elusive. On the one hand, he describes it as something we all experience: When asking myself the question of "Who am I?" I certainly don't do so from a desire to find out my own name, gender, age, etc. Instead, I am searching for something "behind" these personal details — something unique that does not and cannot belong to anyone else. And yet, says Ryle, we cannot put our finger on what this "I" stands for. It is like my shadow — always a pace ahead of me, leaving open what the next step will be. The shadow of oneself will never wait to be jumped on; it evades capture, and yet is never very far ahead.[3]

On the other hand, Ryle dismantles this experience as something spurious. He actually used the term "category mistake," in which things belonging to a particular category are presented as if they belong to a different category. One of the examples he mentions is the following.[4] Someone asks at the end of a tour through all the facilities and laboratories of Oxford University: "Now, where is the University?" That's a strange question. The error this person makes is presuming that the University is part of the category "buildings" rather than part of an "institution." Obviously, the University itself is not another building. According to Ryle, something similar holds for concepts such as "I" and "self." You cannot ask, after having seen all the parts of a body, "And where is I?" Therefore, according to Ryle, separating "I" and "self" from the body is a category mistake, which leads to some kind of "a ghost in the machine" theory.[5]

Is this the end of "self"? No matter what Ryle claims, there is definitely something enigmatic about this situation that no one can talk away or take away from us. Although Oxford University cannot be "seen," it is

2 Gilbert Ryle, *The Concept of Mind* (Harmondsworth; Penguin Books, 1980 [1949]), 186.
3 Ibid., 178.
4 Ibid., 18–19.
5 Ibid., 17.

nevertheless a very real entity. True, it is not one of its visible buildings, yet it is the "unseen" entity behind all those university buildings taken together. If there were no Oxford University, those buildings would not be connected the way they are, or they might not even exist. So there is something very real, though not physical, about the University itself that unites all the buildings one can see and visit. What we "see" can sometimes be best explained by what we do not "see." In other words, one needs to "connect the dots" to see the invisible University through or behind all its visible buildings. Without the invisible University those visible buildings would not even exist.

Something similar could be said about "self" — it is something invisible behind what is visible. Still, the invisible "self" has been under another attack for quite a while. The problem became more acute after the French philosopher René Descartes distinguished "physical substance" [*res extensa*], which can be measured and divided, from "thinking substance" [*res cogitans*], which is not extended and not cleavable.[6] The body clearly belongs to the former category, whereas the mind was viewed by Descartes as something non-physical, something lacking shape, size, and location. In his famous summation, "I think, therefore I am," this "I" was considered a non-physical thing, an "I-thing," existing independently of the physical brain, yet maintaining some kind of causal interaction with the physical connections of nerves and muscles. Strangely enough, Descartes thought that perhaps the pineal gland was the interaction point between the two.

So the question arises whether this "I-thing" could be located somewhere in the body. A popular theory in vogue for a long time was that the body works with something like an "inner observer" or "mental agent" — something like a "man in the machine." Consider watching a movie: Viewers see the images as something separate from themselves projected on the screen. Could we perhaps explain this by supposing that the light from the screen forms images on the retina that are transferred next to the visual cortex, where they are then scanned by an "inner observer" located in the brain behind the cortex? Or when people give commands to their body, could it possibly be that some "mental agent" is steering the body like a piano player would play the piano?

6 René Descartes, *Discourse on Method and the Meditations*, trans. F. E. Sutcliff (Harmondsworth: Penguin Books, 1968), Second Meditation, part 1.

No, not really. If this theory were correct, we would end up with "an observer in an observer" or "a piano player inside a piano player." We would be in for an infinite regress! This is sometimes called the *homunculus* (little man) theory—that is, explaining a phenomenon in terms of the very phenomenon that it is supposed to explain. We cannot explain "man" by assuming a "man in the man-machine." Even the brain does not harbor any detectable agent who plays the "piano" of the body. Of course it is well known that parts of the brain "mirror," or represent, certain parts of the body; and since scientists always search for material causes, they have also been searching for a part of the brain that mirrors and regulates the brain itself. But there can be no such physical brain part that is in charge of the brain; the brain cannot have a part that mirrors itself.

In order to avoid the "homunculus fallacy," its critics reject the idea that self-awareness comes from some inner little being (*homunculus*). So they let the brain do all the work on its own. Some even keep searching for a part of the brain that allows us to be self-aware. But as Max Bennett and Peter Hacker have argued, the homunculus fallacy keeps coming back here in another form.[7] Now the homunculus is no longer an inner being, but a brain part that is supposed to "process information," "map the world," "construct a picture" of reality, and so on. Oddly enough, this brings the homunculus back in a hidden way, for all these latter expressions can only be understood because they describe processes with which our "selves" are already familiar. To describe the resulting form of "brain-science" as an explanation of self-awareness is therefore questionable, as it merely reads back into the explanation the feature that needs to be explained. It creates the unjustified impression that self-awareness is a material feature of the brain (see chapter 11).

It may actually be very frustrating to locate "I" and "self" and "mind" in one particular bodily organ, the brain. Yet that has been tried many times by what we could call the "mind-brain-equalizers." Not surprisingly, Darwin is one of them. Instead of speaking of "the *brain* of a monkey" he refers to "the *mind* of a monkey."[8] However, mind-brain-equalizers run repeatedly into trouble. They would for instance stumble upon the following problem: When neuroscientists study the brain, we should

7 Maxwell R. Bennett and Peter M.S. Hacker, *Philosophical Foundations of Neuroscience* (Oxford: Blackwell, 2003).
8 Darwin's letter to W. Graham, 1881.

wonder how the brain of these scientists could ever be able to study itself. It is hard to see how that would be possible, for it would be like a camera taking a picture of itself. It is much easier to see how the *mind* can study the brain, but it is hard to see how the *brain* could ever study itself.

Indeed, the brain can be studied by scientists, but it is the mind of the scientists, not the brain itself, that does the studying. The physical world can never be studied by something purely physical, any more than DNA could ever discover DNA, or neurons could ever discover neurons. To solve this problem it might be helpful to make a distinction between "subject" and "object." An object cannot study itself; only a subject can study an object. The subject that studies the brain must be "more" than the object it studies, the brain itself, in the same way as Watson and Crick must have been "more" than the DNA they discovered and studied. To put it in more philosophical terms, the *knowing subject* must be "more" than the *known object*. When science studies the brain as an object, such can only be done thanks to the mind of a subject, the scientist. When studying the human brain as an object of science, a scientist needs the human mind as the subject of science—for without the human mind, with its faculties of language and rationality, there would be no science at all.

In other words, it is very doubtful whether these faculties come with the brain, as many scientists assume; instead, it makes more sense to attach these faculties to the mind rather than to the brain, and to attach the mind, in turn, to "I" and "self." It is because a person is a unity of body and soul, of mind and brain, that the mind can study the brain. Interestingly enough, the pioneer of quantum physics, Max Planck, said something similar: "Science cannot solve the ultimate mystery of nature. And that is because, in the last analysis, we ourselves are a part of the mystery that we are trying to solve."[9]

To link this discussion back to what we said earlier about "I" and "self," we could make a distinction between "I-as-an-object" and "I-as-a-subject." I-as-a-subject (I-now, my "self") can reflect on I-as-an-object (I-past). I can remember my past because I-now is always "more" than and "ahead" of I-past. As a subject, I may investigate I-as-an-object and then realize, for instance, that I-as-an-object made a mistake. In contrast, I-now is

9 In Ken Wilbur (ed.), *Quantum Questions* (Boston: New Science Library, 1984), 153.

never open to investigation because its future possibilities are beyond its current actualities — and therefore, I-now will always be a pace ahead of I-past. No matter what I think when I am thinking, it is always I-now who is thinking something, including thinking about I-past.

We can objectivize anything we want, but we cannot objectivize I-as-a-subject, since that is the very one that has the capacity to objectivize. Put differently, "I-as-a-subject" is not an object like other objects in this world, but it is their very origin, for without it, there would be no objects of knowledge comprehensible to us. That is why I can never blame my glands or my animal ancestry for what "I" do wrong, because I-now, my "self," is always a pace ahead of I-past, including my glands. "I-as-an-object" is my body, which may appear in the iron grip of determinism, but "I-as-a-subject" is my "self," free to take a next step, a step ahead of any physical determinism, by acting as a new cause with a new effect. I am always more than my genetic code, for I could have an identical twin, with the same genetic code, but I would never be him or her.

Some philosophers speak rather of a "third-person ontology" versus a "first-person ontology."[10] Phenomena in the brain have a third-person ontology, whereas phenomena of the mind have a "first-person ontology," being essentially subjective or "private," directly accessible only to the subject who has such mental experiences. No matter how we look at it, there seems to exist in this Universe a dualism of properties — material versus immaterial, neural versus mental, objective versus subjective, brain versus mind. We can tell them apart without setting them apart. The mind has distinct features — such as intimacy, privacy, first person perspective, and unity of conscious experience — which cannot be found in the brain and its overt, public, third-person ontology.

Apparently it is not so easy, as some think, to get rid of the "self" behind self-awareness. The mystery remains; we cannot detect any "piano player," yet we do hear his or her "music." Because the mind does not occur on the physical map of the body, it may seem as if it is nowhere in the chain of bodily activities, and yet it is the "soul" of it all and pervades the entire body. Contrast, for instance, a blinking eye with a

---

10 E.g., John Searle, *Mind, Language and Society* (London: Weidenfield & Nicolson, 1999), 42.

winking eye. Blinking is caused by a cascade of neural activities, but winking is more than that—it includes a mental *intention* of winking at someone. In contrast to a blinking eye, there is a "self" or an "I" behind a winking eye.

This takes us back again to that inescapable "systematic elusiveness of 'I'" which is inaccessible to anyone but myself. Put differently, the world of my "self" is a private world. Mental events happen in my private world that I have exclusively to myself, a world to which no one else has direct access. Whereas the world of my body is public and accessible to others, the world of my "self" is private. It's through introspection (and retrospection) that I have access to my own private world in a way no one else has. Even brain scans have no access to my private world as I myself do; all they could pick up is "brain waves," but never my thoughts, for those fail to show up on pictures and scans.

It is here that we come in touch with this astonishing capability of my "self": the faculty of reflecting on my "self." I may notice for instance that I am clumsy; I may even notice that I am laughing at myself for being clumsy; and then I may decide to tell others that I noticed how I was laughing at my own clumsiness. This is in fact an iterative or recursive process of self-reflection, making "I" act like my own shadow—I-now reflecting on I-past. I can never get away from my "shadow" in the same way as I can get away from someone else's shadow. The fact that I can point to myself is rather peculiar, though. A missile, for example, cannot be its own target; the index-finger cannot point at itself. But I myself can certainly be my own target. I can think about myself, correct my own actions, comment on my own actions, even revise the comments on my own actions. This is the mysterious power of "self." However it is always my very "I" (I-now) who is doing and thinking this! There is no self-reflection without self-awareness.

Self-awareness is a faculty that operates under the power of the mind with its faculties of language, rationality, and morality. It is very hard, arguably impossible, to describe this power in purely material terms derived from physics, biology, neurology, or genetics. That makes it very questionable whether our understanding of the world can be entirely attributable to the material mechanism of the brain. The philosopher Michael Augros uses the following analogy: You cannot count what you are seeing without using your eyes, but that does not mean your eyes are doing the counting.[11] Similarly, it is clear that we cannot understand anything without using our brains, but it does not follow that our brains are doing the understanding. It is "I" who is doing the understanding, no matter how "elusive" this "I" may be in the eyes of some philosophers or scientists. So this poses the question as to where this "I" of self-awareness might ultimately come from. Could we have inherited it from the pre-human animal world?

THE MIRROR TEST

On a visit to London Zoo in March 1838, Darwin stepped into a cage with an orangutan named Jenny, and marveled as she played with a mirror. He wondered what exactly the ape made of the image of herself. Darwin was always good at placing himself in the "shoes" of an animal. We all seem to do that too at times. Whenever we try to assign thoughts, intentions, and expectations to animals, we are actually picturing ourselves in the "shoes" of an animal and proclaiming "on behalf of" the animal what we would think or say were we ourselves to bark at a cat in a tree, and so on. It is probably just a way of speaking, of course, but some seem to take this metaphor quite literally.

Here is what the psychologist Gordon G. Gallup did. In 1970 he devised the so-called mirror test to examine self-awareness in animals by studying whether an animal can recognize its own reflection in a mirror as an image of itself. Through a hole in the wall, Gallup observed how chimps first treated their reflection as if it was another chimp. But over time, his chimps started using the mirror to explore their own bodies, using it to look at the inside of their mouths, to make faces at the mirror. But Gallup also realized that this presumed evidence for self-awareness was not very compelling. So he figured out a better technique for his mirror test. He anesthetized the chimps, then painted one eyebrow ridge and the opposite ear tip with a red dye that the chimps could see through a mirror but wouldn't be able to feel or smell. Once awakened, after looking in the mirror, the chimps used their fingers to touch the red dot on their forehead and after touching the red dot they would even smell their fingertips.

---

11 Michael Augros, *Who Designed the Designer? A Rediscovered Path to God's Existence* (San Francisco: Ignatius Press, 2015), 123.

As far as Gallup was concerned, this was the proof he needed. He exclaimed, "the first experimental demonstration of a self-concept in a subhuman form." He added, "It didn't require any statistics. There it was. Bingo."[12] He concluded: "Animals that can recognize themselves in mirrors can conceive of themselves." Then he further added: "It's not the ability to recognize yourself in a mirror that is important.... It's what that says about your ability to conceive of yourself in the first place." Gallup's convictions became only stronger over time.[13]

The mirror test has been repeated many times since, and on more and more animals. Animals which have "passed" the mirror test are chimpanzees, bonobos, orangutans, dolphins, elephants, and possibly pigeons. Surprisingly, gorillas have not passed the test, except for one specific gorilla, Koko; this might be because gorillas consider eye contact an aggressive gesture and normally try to avoid looking each other in the face. Yet, the general feeling remains that the mirror test has proven self-awareness in at least some non-human animals — they must have a "self." Since having a "self" presumably comes with personhood and human rights, the mirror test has already opened a new territory for the Civil Rights Movement. Some gradualists even argue that passing the mirror-test indicates a level of self-awareness that makes it unethical to keep animals in captivity. Next we should be waiting for these animals to express their desire to take "selfies."

But let's not jump to conclusions too quickly. Even if certain animals passed the mirror test, does that mean they also passed the self-awareness test? Not necessarily so. The mirror test has remained very controversial — not among gradualists, of course, but among their critics. Here are several reasons for their criticism.

First of all, mirror self-recognition is at least partially a learned behavior. When first exposed to a mirror, most apes react to their reflection as they would to another member of the same species; but eventually they may learn to recognize their mirror image as a reflection of their own bodies. Even Gallup noticed how chimps first treated their reflection as though it was another chimp. But only over time did they start using it

---

12 G.G. Gallup, "Chimpanzees: self-recognition," *Science*, 1970, 167 (3914): 86–87.
13 Gallup Jr., et al., "The Mirror Test," in M. Bekoff, et al. (eds.), *The Cognitive Animal: Empirical and Theoretical Perspectives on Animal Cognition* (Chicago: University of Chicago Press, 2002), 325–33.

to explore their own bodies. Usually, learned behavior works with associations, not concepts such as "self." It is a matter of making connections between particular external stimuli and their associated signals without using any mind-dependent concepts (see chapter 7). The fact that birds, for instance, can clean their own feathers doesn't mean they have the capacity of self-awareness.

Second, one could argue that passing the mirror test is merely evidence of good intelligent abilities, possibly based more on face-recognition than a sense of "self." Dogs, for instance, perform poorly on the mirror test, but they do seem to be able to recognize their own scent. We discussed earlier that classification and categorization are rather common and basic in the animal world. If a prey animal can identify a lion as a "predator" it can probably also distinguish its own body from other bodies, without necessarily having a "self." If animals are very well able to classify, distinguish, and recognize faces and scents, then they might also be able to classify a face or scent as coming with one particular body, their own.

Third, the mirror test may indeed show us that some animals can recognize their own body. However, awareness of their own body is not necessarily self-awareness. Animals that regularly drink from water surfaces should be accustomed to seeing a reflection of their own bodies in the water anyway. But do they have a sense of "self," a sense of past and future, of knowing that they exist in a particular time and place? It is very doubtful, for what mirrors mirror is a body, not a "self." Animals won't just go to the water to look at themselves in the "mirror," whereas humans tend to revisit their mirrors many times a day. In other words, animals are able to use "perceptual abstraction"—the difference between "this body" and "those bodies"—but not necessarily "conceptual abstraction"—the difference between "self" and "others" (see chapter 7).

Fourth, the mirror test only tests the ability of association. Daniel Povinelli, now at the University of Louisiana-Lafayette, has become one of Gallup's most outspoken critics.[14] He has come to believe that a chimp doesn't need to have an integrated sense of "self" in order to pass the

14   Daniel J. Povinelli, et al., "Toward a science of other minds: Escaping the argument by analogy," *Cognitive Science*, 2000, 24 (3): 509–41.

mirror test. Instead, the chimp needs only to notice that the body in the mirror looks and moves the same way as its own body, and then makes the connection that if there's a spot on the body in the mirror, there could also be a spot on its own body. That may just be a matter of learning and association. In a similar way, when we use the mirror to direct our toothbrush, we are not thinking deeply about our "self." The mirror is just a helpful tool like all other tools. So why not for animals?

Fifth, the mirror test could actually be considered very "biased." Povinelli calls the mirror approach a case of "folk psychology"[15] — making unscientific inferences based on our own human experiences. In his own words: "Why devote so much energy to trying to determine if chimpanzees have a human-like theory of mind? Why not try to figure out what makes them chimpanzees, instead?" If concepts such as "self" and "mind" are indeed a unique specialization of the human species, as we argued before, then all efforts to apply them also to the non-human animal world are actually biased examples of profound anthropomorphism.

Finally, the mirror test may tell us much more about ourselves than about the animal world. When we see apes recognizing "themselves" in a mirror, we hope — or perhaps — fear, to see ourselves in them, but that doesn't say much about an animal's "self." It is exactly because animals do *not* have a "self" that they can't be altruistic, unselfish, or even selfish, for those terms belong to a moral context (see chapter 8). When animals fight with their rivals for food, they may indeed seem to know the difference between "them" and "me," but "me" does not automatically imply a "self," let alone self-awareness.

Given the above reasons, we may conclude that, although the mirror test keeps ranking high on the agenda of Neo-Darwinian evolutionists, it doesn't entitle us to equate recognition of one's own body with self-awareness. Without a "self" there is no self-reflection, no self-expression, and no self-determination. In spite of claiming continuity in evolution, the awareness of "self" still seems to be a unique human feature, similar to what we found out earlier about the faculties of language, rationality, and morality. We are the only beings in the Universe who change in order to be more what we are.

---

15   Daniel J. Povinelli and Jennifer Vonk, "We don't need a microscope to explore the chimpanzee's mind," *Mind and Language*, 2004, 19 (1): 1–28.

AT THE DAWN OF SELF

How did the human "self" emerge? We don't know much — in fact virtually nothing — about the brain wiring behind this faculty of self-reflection, let alone its genetic basis. So where could we start? Gallup, the father of the mirror test, was probably right on at least one issue: strong self-awareness may also entail *death*-awareness. As he puts it, "The next step, it seems to me logically, is to confront and eventually grapple with the inevitability of your own individual demise." And, "Death awareness is the price we pay for self-awareness."[16] This means that we realize we can lose our "self" when we die.

This may open a window to the question of *when* the human "self" emerged. If there is any indication of death-awareness, we may assume there is also self-awareness. The problem is, though, that we could never find this out by merely looking at skulls and skeletons, for awareness does not fossilize. In other words, the question whether skeletal remains — for instance of Neanderthals — came from humans with a "self" has nothing to do with their biological "looks," but depends rather on whether they were beings with the faculties of language, rationality, morality, and self-awareness — and therefore, death-awareness.

Paleoanthropologists have no way of spotting death-awareness, other than going by burial practices, which are an indication of death-awareness — and therefore, of self-awareness. They consider evidence of burial rituals and the presence of grave goods an important indicator of being human because these may signify self-awareness as well as a concern for the dead that transcends daily life. Burial of the dead with material goods indicates a belief in an afterlife, for the goods were seen as useful to the deceased in their future lives where their selves could go on.

However, even burials are hard to diagnose. There are serious questions for instance as to whether Neanderthals practiced burial rituals. One of the main critics is Robert Gargett, and to some degree Paul Pettitt.[17] Gargett maintains that these presumed intentional burials had likely been caused by natural processes and phenomena.[18] It has also been argued that

16   G.G. Gallup, Jr., "Self-awareness and the evolution of social intelligence," *Behavioural Processes*, 1998, 42, 239–47.

17   P.B. Pettitt, *The Palaeolithic Origins of Human Burial* (New York: Routledge, 2011).

18   Robert H. Gargett, et al., "On the Evidence for Neanderthal Burial," *Current Anthropology*, 1989, 30 (3): 322–30.

nothing can convincingly be characterized as Neanderthal "grave goods." It should be emphasized that most Neanderthal remains (and there are hundreds) have not been found in graves — and even the few graves found are controversial, regardless of what gradualists claim. There is no clear evidence for ritual behavior of any kind — no grave offerings, no ceremonial fires. The problem is that while burial is intentional, the intention may simply have been to bury the individual to prevent predators from being attracted to the area or to deal with contamination. How different this is when we deal with remains of early modern humans of the species *Homo sapiens*! The earliest, virtually undisputed human burial dates back some 80,000 years, in the Skhul Cave at Qafzeh, Israel.[19] It is very likely we are dealing here with human beings who had a sense of "self."

Having a sense of "self" and of "death" is missing when we look at the non-human animal world. Take for instance the observation of female apes that keep carrying their dead newborns around for quite a while, without having any idea of what has happened until at last they give up and drop the dead remains of their newborn. In Guinea, West Africa, chimpanzee mothers were seen in nature carrying and grooming their offspring's lifeless bodies for up to sixty-eight days.[20] By the time the corpses were finally abandoned, the bodies had developed an intense smell of decay. Ironically, evolutionists gave this observation a peculiar twist: these, they say, must be sixty-eight days of actually mourning the dead! Only human beings could come up with such an explanation. Arguably, this sixty-eight-day period may not be a time of "mourning" but rather a time of "not-knowing." Besides, we should ask these evolutionists where the burial and the grave are once this presumed period of "mourning" had ended. As we said earlier, for animals there is no "tomorrow," for that requires having the concept of "tomorrow"; that's why animals cannot think about the possibility that someday there may not be a "tomorrow" in their lives.

In other words, it's hard to see how our faculty of self-awareness and death-awareness could have ever come from our pre-human ancestors.

---

19 E. Trinkaus, "Femoral neck-shaft angles of the Qafzeh-Skhul early modern humans, and activity levels among immature near eastern Middle Paleolithic hominids," *Journal of Human Evolution*, 1993.

20 D. Biro, et al., "Chimpanzee mothers at Bossou, Guinea carry the mummified remains of their dead infants," *Current Biology*, 2010, 20 (8).

Legendary biologist Theodosius Dobzhansky could not have said it better: "A being who knows that he will die, arose from ancestors who did not know."[21] And, "There is no indication that individuals of any species other than man know that they will inevitably die."[22] Only humans can ask, "Why me?" or "Why do I have to die?" So there is no other conclusion than this: How different was life, and death, when the first human beings arrived!

Once the first humans had arrived, not only do we find evidence of elaborate burials for the first time in prehistory, but evidence also of aesthetic expressions. Findings in prehistoric graves suggest that decoration and art were an integral part of the lives and societies of the people who made them. Which means that these humans were not only masters of self-awareness but also of self-expression. We mentioned already the earliest symbolic artifact found thus far: a piece of ochre with a crosshatch design carved in it, dated to about 80,000 years ago in the Blombos Cave in South Africa (see chapter 6). The makers of this artifact showed clear signs of symbolic behavior—and therefore, most likely, of language and rationality—before their migration to Europe. But now we also see a new dimension to self-awareness: self-expression.

Human beings have this astounding capacity of expressing themselves in various forms of art—in literature, in paintings, in music. We find these expressions as far back as the early history of humankind. Of all animals in prehistory, only humans left behind stones with inscriptions; only humans created paintings in caves, etc. As a matter of fact we have found astonishing art in prehistoric caves. Understandably, some gradualists considered such paintings fake for a while, because they could not accept such a high level of self-expression so early in human history—again based on their ideology of gradualness. A simple response to them would be: Even a cave man can do it! Some of the first Cro-Magnon sites, dating from well over 30,000 years ago, have even yielded evidence for music: multi-holed bone flutes capable of producing a remarkable complexity of sound.

Yet the novelty of self-expression is hard to accept for gradualists. They have been feverously searching for pre-human features in the animal

---

21 Theodosius Dobzhansky, "Changing Man," *Science*, 1967, 155, 68.
22 Theodosius Dobzhansky, a.o., *Evolution* (San Francisco: W.H. Freeman, 1977), 454.

world, so even self-expression has become their target. They tried, for instance, to let apes create their own paintings.[23] However, the colorful mess apes produced under pressure is far from any representational kind of art that we know of among humans. Chesterton called art "the signature of man."[24] Gradualists also like to point at the beautiful tunes songbirds can generate, but those are mere repetitions of the same, old tune — certainly not music. Non-human animals just seem to lack the faculty of self-expression. The most likely explanation for this missing feature is that they have no real "self."

Let us not forget that only humans are able to laugh and cry, which is another sign of self-expression. Crocodiles may shed tears while devouring their prey, but we can be certain it is not because of remorse. Chesterton says about man, "Alone among the animals, he is shaken with the beautiful madness called laughter."[25] Why is it that tears of laughing and crying are so unique to human beings? The answer is again that only humans have self-awareness, rationality, and morality — and can therefore express themselves in this way.

If gradualists tell us, in response, that we only developed self-expression by mimicking others, we should ask them why our pets did not grab their opportunities, then, to become act- and look-alikes of their owners. Apparently we must be overlooking something essential. Isn't it amazing how the ideology of gradualists with their slogan of "Apes 'R' Us" has nothing new to offer to an emerging humanity — and yet, a great deal seemed to be ground-breaking when humanity emerged. Just ask yourself why some apes are still swinging from trees, while humans are able to walk on the moon — something apes can't even think of doing. Chesterton said it well: "the more we really look at man as an animal, the less he will look like one."[26]

Thanks to self-awareness, a human being can act like a subject studying other objects as if they were subjects as well. We are good at imagining what the world would look like through the eyes of others. However, this ability of reading someone's mind is unique to the human mind.

---

23   Desmond Morris, *The Biology of Art: A Study of the Picture-Making Behaviour of the Great Apes and Its Relationship to Human Art* (London: Methuen Publishing, 1963).
24   G.K. Chesterton, *The Everlasting Man*, part 1, chapter 1, "The Man in the Cave."
25   Ibid.
26   Ibid.

No dog will ever come up with the Cartesian thought "I bark, therefore I am—just like you." So we shouldn't be surprised that we have never seen chimps winking at their trainer. What does a "wink" add to a "blink" then? The answer is simple: My wink tells you to read my mind as kidding. Humans are good at picturing themselves in someone else's shoes, even in the "shoes" of a chimp, but chimps cannot picture themselves as playing a trick on someone else. For animals, that's just too elusive. There are reports that chimps can fool or deceive each other, but we should always wonder if this is really intentional deception, or just a tactic learned by trial and error. Are they really "reading minds"? It is, again, our unique human mind that interprets this as deception—for that's how we ourselves would do it!

Here is what we should conclude from all this: No animals seem to pass the "self-awareness test." There appears to be another deep divide between them and us. The best we can say right now is that there is a deep self-awareness divide between the pre-human and human world. There is no way of telling whether this leap can be fully explained in terms of genes and mutations. Obviously, the faculty of self-awareness requires a body with the "right" features. The "right" mutations were *necessary* to achieve this. But whether those mutations were also *sufficient* for self-awareness is impossible to tell ahead of time. Those who think this divide can eventually be completely explained by science must realize that this cannot be more than a conviction—more of a program than an achievement. There is no scientific way of corroborating such a conviction—it is worth as much as the opposite claim that a scientific explanation will never be fully possible.

10

The Religion Divide

THERE IS ANOTHER ASPECT OF HUMAN SELF-AWARE-
ness that we need to bring up: It is the faculty of self-transcendence.
Religion presupposes some form of self-transcendence—knowing that
there is Someone who transcends my "self"—who is greater than me,
who created me, and who sustains me. Religion is also a way of coping
with our knowledge that death is inevitable. Religion diminishes the hurt
of death's certainty and the pain of losing a loved one with the prospect
of life-after-death. Humanity has always known this—intuitively, if you
will—from its very beginning.

SELF-TRANSCENDENCE

When I say "I am only human," I am not comparing myself with
something "below" me (such as a cat, dog, or chimp); I am comparing
myself with Someone who is "above" me and transcends me. When I call
myself "only human," I am actually comparing myself with a Person who
does not have the limitations I myself experience. In some mysterious
way I am reaching out into the realm of the Absolute, far beyond myself.
In doing so, the "finite" catches a glimpse of the Infinite. Apparently, the
"finite" human mind is able to catch no more than a "glimpse" of the
Infinite. To use a poor analogy: If I can count from one to ten, or from
one to thousand, then I can also count from one to infinity, at least in
theory. This faculty of the human mind for the infinite is rightly called
self-transcendence—referring to the transcendence of an Infinite Being
far beyond our own selves. Belief in the transcendent is at the heart
of all religions. Of course, I cannot transcend myself on my own, but
because I myself was made in the image of God, I perceive more than
myself whenever I perceive myself completely.

It is man's drive for self-transcendence that explains why belief in God
as a Transcendent Being is so universal, and yet so unique, in human
history. Belief in the transcendent is at the heart of all religions; it unites
the orthodox forms of Judaism, Christianity, Islam, Buddhism, and even
Hinduism. The rising number of atheists in modern society seems to belie

this, but we need to acknowledge that even among philosophers—the professional skeptics among us—there has been little controversy about the existence of some absolute or infinite being. As the late philosopher Fr. Joseph Bochenski, O.P. put it a few decades ago:

> The fact that there is such a being [an absolute, Infinite being] is the common conviction of Plato, Aristotle, Plotinus, St. Thomas, Descartes, Spinoza, Leibniz, Kant, Hegel, Whitehead—and even of dialectical materialists, the official present and past party philosophers of communism, if these smaller minds are, in any way, comparable with the greater ones. Although all these [communist] philosophers vehemently deny the existence of the Judeo-Christian God, they claim at the same time that the world is infinite, eternal, boundless, absolute. And what is more, their attitude is in many ways distinctively religious.[1]

These philosophers in one way or another all claim that God's existence is not a matter of opinions but a *factual* issue. It is a fact that God either exists or does not exist. All monotheistic religions claim there actually *is* a Person who transcends me. But at the same time they assert that God has to be radically different from everything we see around us, including each one of us.

Thomas Aquinas is famous for putting this thought of an absolute, infinite Being—the very core of religion—into philosophical terms. His terminology may seem foreign at first, but it has withstood the test of many centuries. His reasoning starts as follows: we *receive* our existence; we are *contingent* and could easily not have been; we don't have to exist, but because we do exist, we can ask for the cause of our existence—God. God alone is the act of existing itself. God alone is existence by nature. So God is a necessary and eternal Being who did not come into existence but always has been. In contrast, all other creatures are not God precisely because their act of existing has been received from God, who alone is self-existing. Each creature exists because God, who is existence itself,

---

1 Joseph Bochenksi, *The Road to Understanding: More than Dreamt of in Your Philosophy* (North Andover, MA: Genesis Publishing Company, 1996), 116.

holds that creature in existence at every time and place. If God did cease to hold each one of us in existence, we would simply disappear—we would be an-nihil-ated, back to what we came from, nothing (*nihil*).

Based on this foundation, Thomas Aquinas developed his famous distinction between "Primary Cause" and "secondary causes," which we mentioned already (see chapter 2). He calls God a "Primary Cause," and all the causes that nature deals with he considered "secondary causes." The physical causality of nature reigns "inside" the Universe, linking causes together in a chain of secondary causes. God, on the other hand, reigns from "outside" the Universe as a Primary Cause or First Cause, thus providing some sort of "point of suspension" for the chain of secondary causes itself, so to speak. A thing can cause *other* things to change but it cannot be the cause of its *own* existence. What is so special about the Primary Cause is that it needs *no* cause. And what is so special about secondary causes is that they *do* need a cause. The difference between God and the world is not that one has an explanation and the other lacks it, but rather that one is self-explanatory while the other is not.

To explain this important distinction, I would like to introduce an example that the philosopher Edward Feser uses: a cup of coffee sitting on someone's desk.[2] The cup has no capacity on its own to be a few feet from the ground; it will be there only if something else, such as my desk, holds it up. But the desk in turn has no power of its own to hold up the cup. The desk too would fall to the earth unless the floor held it aloft. And the floor, for that matter, could hold up the desk only because it is itself being held up by the house's foundation, and the foundation by the earth, and the earth by the structure of the universe. Something similar could be said about a book sitting on the shelf of a bookcase, and many other examples. All these "intermediaries" keep each other in tow. However, none of these things could hold up anything at all unless there were something that holds them up without having to be held up itself. That ultimate "something" is the Primary Cause that keeps everything else in existence. Without the Primary Cause, nothing in this series of causes would really be explained, for without the Primary Cause holding up all the secondary causes, the entire series of causes would collapse, or

2   Edward Feser, *Five Proofs of the Existence of God* (San Francisco: Ignatius Press, 2017), 22f.

not even exist. This Primary Cause has the power to produce its effects without being caused by something else. It has *inherent* causal power, while the secondary causes have only *derived* causal power.

It is the secondary causes that we are most familiar with — "like causes having like effects." Science deals with this kind of causality. It is the causality that reigns "inside" the Universe, linking causes together in a chain of secondary causes. How can we explain this chain of causes? Circular causation is obviously out of the question. The philosopher Michael Augros gives a few simple examples: "You can't be your own father. You can't give your own existence to yourself or receive it from yourself."[3] Infinite regression won't work either. True, a sequence of events in time may be able to go infinitely forward through the future and back through the past, but the problem would be that time itself is still in need of some kind of cause, as is this very series of causes itself. As Edward Feser puts it, an infinite series of causes can no more get you a real something than an infinite series of IOUs can give you real money — at some point, these IOUs have to be backed up by real money.[4] In other words, the chain of causes needs to be hooked on to something else, so it doesn't hang from nowhere or just float in the air.

Besides, no part of the chain can do any causing unless it first exists. In other words, the need for causes must come to an end: there must be or have been a cause that is not itself in need of a cause — a Primary Cause. The Primary Cause does not need a cause — not because it is an arbitrary exception to the general rule, but because it has always existed and did not come into existence, and therefore doesn't require a cause. Even in case of an infinite series of contingent things, each caused by a previous one, there is still the question why the series itself exists at all, for the series is just as contingent as its individual contingent elements. That's where we need a necessary, self-explanatory, and uncaused Primary Cause for all secondary causes. There will have to be some cause *outside* the series, which imparts causal power to all of the series' members. Nothing less than such a necessary Being could possibly terminate the regress of explanation. If we deny this, then we merely pass "the explanatory buck rather than explaining anything."[5]

3   Michael Augros, *Who Designed the Designer?*, 33.
4   Edward Feser, *Five Proofs of the Existence of God*, 133.
5   Ibid., 157.

What Aquinas does in all of this is to safeguard God's *transcendence*. Because of God's transcendence, God is not a deity like Jupiter or Zeus—not a being stronger than other beings or superior to all other beings, yet acting like all other beings. Instead, God is the very source of all being—the Absolute Ground of all that happens to exist. This Primary Cause is un-caused, not even self-caused, but the Source of all being; not a super-being among other beings, who acts like other beings, but Absolute Being; not a cause prior to and larger than other causes, but a Primary Cause; not a power stronger than and superior to all other powers, raised to the power of a zillion, but instead an Infinite Power; not a super-cause among other causes, but a Power "above" and "beyond" all secondary causes. The core message is this: When the transcendence of the First Cause is overlooked, the existence of secondary causes becomes either an enigma or a contradiction. God is not a rival or contender for created causes, but rather the One who makes all secondary causes to be their own causes. Even Plato and Aristotle knew that (the real) God is unlike Jupiter or Zeus.

Back to our concept of self-transcendence. Self-transcendence does not mean that we project our "self" into someone who is like our own "self" raised to the power of a zillion—which is something Sigmund Freud would proclaim. The perfections we find in God are not super-traits. God's power is not a super-power, but the Source and Origin that all human power depends on and derives from; other powers are not "next" to God but "under" God. We ourselves would not have any power if God did not give us power. The same holds for God's omniscience; it is not the sum total of all human knowledge, but surpasses it in an infinite way. And this holds also for God's omnipresence: it is not the presence of all individual creatures compounded together, for it is of an entirely different magnitude and scale. If God is indeed omnipresent, "everywhere," God may seem to be "no-where," but God's omnipresence only makes for God's *seeming* absence. As they say about fish: The last thing a fish would ever discover is water. Something similar holds for us discovering God's existence.

The conclusion of all this is that there must be someone behind and beyond all we see and experience—and this "someone" is the Creator of Heaven and Earth. Only in the above sense can it be said that God is all-powerful, all-knowing, and all-present. He is infinitely greater than all his works. Put differently, with the words of Michael Augros, "[t]he step from man to God is infinitely greater than the step from beast to man. A toad would stand a better chance of fully comprehending a man... than a man would have of comprehending god."[6]

Needless to say, the God of the "Primary Cause" is in essence the "God of philosophers and scholars"—and not, as Blaise Pascal put it in his *Memorial*, the "God of Abraham, the God of Isaac, and the God of Jacob." Thomas Aquinas made it very clear: "We must believe that a god exists, which is clear by reason." What Thomas means by this statement is that the *reason* of rationality leads us to the existence of a Primary Cause. This does not mean that reason can fully fathom God; philosophical knowledge is limited to knowing *that* God is, more so than what or who he is. But this is only part of the story: the *faith* of religion discovers that this very concept refers to the God of our Faith. Reason only tells us that this Universe cannot explain itself, but needs an ultimate non-contingent explanation that can be found only in God. The "rest" has to come from somewhere else: God's revelation in Jesus Christ and the Bible.

To know there must be a God is one thing; to believe in this God and then live and act as if God is real is another. One still needs the grace of Faith in order to embrace what we know through reason. Reason takes us to the door of Faith, but only Faith can take us through that door. It is in the harmony of these two that we come to a deeper understanding of God. Faithful reasoning and reasonable Faith go hand in hand. Obviously, religion is more than a matter of philosophy.

A GENE FOR RELIGION?

What we have discovered so far clearly suggests that religion is unique to humanity and is strongly connected with the human faculty of self-transcendence. Religion is most likely the motivation behind the burial practices we find already in the earliest stages of *Homo sapiens*. This is probably a viewpoint hard for gradualists to accept. They would rather claim that religion must have a genetic basis and could have only emerged through mutation and natural selection of genes found in the pre-human animal world. So it is to be expected that gradualists have been in search of a genetic basis for religion that can be traced back to the pre-human world.

6 Ibid., 140.

This search has become very animated due to a 2005 book by the human geneticist Dean Hamer to which he gave the provocative title *The God Gene: How Faith Is Hardwired into Our Genes.*[7] The god-gene hypothesis was basically *invented* by Hamer, who then claimed he has in fact *discovered* a gene that he decided to call the "god gene." To be more precise, he is talking about a gene for spirituality, which he deceptively dubbed a "god gene." That looks more like an attention-grabbing trick, for his "god gene" is at best a "gene for spirituality."

Hamer theorized that if our sense of spirituality has a genetic basis, those who rank higher in spirituality should share some genetic link which those who ranked lower do not. This raises the question of what he means by "spirituality." This he measured by using a "self-transcendence" scale developed by the psychiatrist Robert Cloninger, in order to quantify how "spiritual" someone is, assuming that spirituality can be quantified by psychometric measurements.[8] What impressed Hamer is that the self-transcendence measure had been shown by classical twin studies to be heritable. (As an aside, those twin studies actually found that specific religious beliefs do *not* have a genetic basis but are transmitted by non-genetic means, such as culture, tradition, and imitation — so they might be inherited without being genetic.)

What about the term "self-transcendence"? It is not taken here in the sense we used it earlier in this chapter. It is not based on belief in God, frequency of prayer, or any other conventional religious practice. Instead it is a word used by psychologists to describe spiritual feelings that are independent of what they call "traditional religion" or "organized religion." In this view, self-transcendent people tend to see everything, including themselves, as part of one great totality. They have a strong sense of "oneness" with people, places, and things. Self-transcendent individuals are also considered "mystical." They are fascinated with things that cannot be explained by science. They are creative but may also be prone to psychosis. In short, they are "spiritual," if you redefine this term to suit your needs.

It has in fact become almost epidemic to replace the term "religion" with "spirituality" — the New Age version of religion, of the "god within."

---

7 Dean Hamer, *The God Gene: How Faith Is Hardwired into Our Genes* (New York: Anchor Books, 2005).

8 C.R. Cloninger, *Feeling Good: The Science of Well-being* (New York: Oxford University Press, 2004).

People who no longer claim they are "religious" still tend to call themselves "spiritual." By this they seem most often to mean a nature-based spirituality that allows them to find something of the divine in forests, lakes, and mountains, where their "spiritual needs" are being met. They find God inside themselves and in nature, not in the pews.

In order to identify some of the specific genes involved in this kind of self-transcendence, Hamer analyzed DNA and personality score data from over a thousand individuals. He asked them to fill out a detailed questionnaire — a standard test called "Temperament and Character Inventory" — including a section asking them to rate their feelings of "absentmindedness, connectedness with nature, belief in extrasensory perception, and other traits." He assumed that their answers would provide a measure of the subjects' affinity for what he called spirituality.

Then he went poking around in their genes to see if he could find the DNA responsible for their differences. With over 21,000 genes and 3.2 billion chemical bases in the human genome, he limited his search for the "spiritual gene" to nine genes known to produce monoamines (brain chemicals that regulate mood and motor control), and then identified one particular gene, VMAT2, as showing a significant correlation with affinity for spirituality. VMAT2 is a gene that codes for a monoamine transporter that plays a key role in regulating the levels of the brain chemicals serotonin, dopamine, and norepinephrine. These monoamine transmitters are in turn thought to play an important role in regulating the brain activities associated with mystic and spiritual experiences.

When he analyzed this gene further, he discovered that those with the nucleic acid cytosine in one particular spot on the gene ranked high in spirituality, whereas those with the nucleic acid adenine in the same spot ranked lower. So he concluded that a single change in a single base in the middle of the "god gene" — at position 33050 of the human genome map, to be precise — seemed directly related to the ability to feel self-transcendence. He even offered an explanation why the "spiritual" allele for this gene would give its carrier a selective advantage: Spiritual individuals are believed to be favored by natural selection because they are provided with an innate sense of optimism, which produces positive effects at either a physical or psychological level. Amazing what natural selection can do! And it all sounds so accurate and precise — so scientific!

What are we to make of all of this? Let me mention first of all that Hamer rushed into print with his book without peer review and without publishing his results in a credible and reputable scientific journal—plus, which is even more serious, his findings have not been replicated. All this probably gives us ample reason to not take his work at face value; or let us say at least that, put nicely, it definitely deserves further scrutiny. Let's do that. What happens when geneticists reduce religion to a set of genes tied to spirituality? There are several serious objections against this kind of approach, and more specifically, Hamer's approach. Let's analyze them step by step.

First a general remark. Hamer has proved himself to be an expert in inventing genes—in 1994 it was a "gay gene,"[9] and now (in 2005) a "god gene." Unfortunately for him, the field of behavioral genetics is littered with failed links between particular genes and behavioral traits. Many of these presumed links were inventions in the mind that never made it to the status of discoveries in reality. In science, discoveries typically start as inventions—usually called hypotheses. However, not all inventions lead to discoveries. To use an analogy, the person who invented or hypothesized "Atlantis" did not discover Atlantis; it remains a legendary island until further notice. The same in science: Most inventions do not lead to discoveries. Yet some scientists think they have made a discovery when all they have in mind is an invention, a speculation. As a consequence we have been flooded with all kinds of new genes. However, a hypothesis is only an invention in the mind until it has been proven to be a discovery in reality as well. As someone said, "Given the fate of Hamer's so-called gay gene, it is strange to see him so impatient to trumpet the discovery of his God gene."[10]

Reason #2. The assumed link between gene VMAT2 and religious experiences is weak. First of all, why did Hamer limit his search to nine genes known to produce monoamines? If a certain allele were indeed the cause of being "spiritual," we should expect that the number of people possessing this allele should at least be proportionate to those who consider themselves spiritual. In addition, all those possessing the "right"

9   Dean Hamer, *The Science of Desire: The Search for the Gay Gene and the Biology of Behavior* (New York: Simon and Schuster, 1994).

10   Carl Zimmer, "Faith-Boosting Genes: A search for the genetic basis of spirituality," *Scientific American*, October, 2004.

allele should have spiritual experiences; otherwise the presence of cytosine cannot be the cause of being or feeling spiritual. Hamer failed to test for any ramifications like these. If our belief in the divine is due to our genetic wiring, how can we not believe in God when the wires are connected? Apparently, there is more to being "hard-wired" for religion. As we said earlier, genes can never produce truths and untruths — certainly not the truths and untruths of religion. The best that genes could ever produce for religion are neural activities that evoke spiritual feelings and emotions. For those skeptical about the legitimacy of religion, that is all they need. But then we are no longer talking religion.

Reason #3. The results of twin studies only speak of some 40% to 50% heritability of spirituality and/or religion.[11] This naturally raises the question what to make of the residual percentages. Twin studies remain very controversial, even when done with identical twins separated by adoption. First of all, similarities between identical twins are not only the result of their identical alleles but also of nearly identical surroundings. Their strong resemblances make it even more likely that others will treat them the same way in life. Moreover, they themselves often strongly desire to become and be more like each other. Hence, we would easily over-estimate the impact of genes.

Second, even if twins become separated at an early age, we need to take into account that these similarities get reinforced the longer it takes before they are separated. Let's not forget that for nine months they shared their mother's womb, including her voice, her hormones, her food, and her emotions. Besides, adoption usually takes place in an environment very similar to the original one, often just around the block or with relatives or friends. So, again, we tend to easily over-estimate the impact of genes. All of this makes research on identical twins, even when they were separated by adoption, rather limited.

Reason #4. Even if it were true that we are genetically hard-wired for religion, what could this possibly mean? Clearly, we are not hard-wired for a specific religion; there are more than 7,000 identified varieties. "Born a Catholic" does not mean always a Catholic. Plenty are the cases of people who in the course of their lives decided to become

11   Tim Spector, *Identically Different: Why We Can Change Our Genes* (New York: The Overlook Press, 2013).

atheists or chose the opposite by leaving atheism behind. But in any of these cases, the changes in religion were almost certainly not caused by genetic changes. Even if some part of spirituality is wired in the brain, the particular forms and practices of religion are still cultural and can be passed from one person to another by learning or imitation, and can be changed by further experiences in life. In a certain sense, Voltaire was right when he said, "All men are born with a nose and ten fingers, but no one is born with a knowledge of God."[12] In that respect, religion is similar to science. No one is genetically determined to join a religious or scientific community (although some might be better genetically equipped or inclined to do so). Members of both communities are in search of the truth, but in order to do so, both require schooling in the form of training and study.

Nevertheless, it is often argued by geneticists that the presence of some form of religion in all cultures of all ages indicates that the idea of God must be preloaded in the genome. However, that comes close to a circular argument: The cultural universality of religion suggests a genetic basis; hence, since there is a genetic basis, religion must be a universal phenomenon in human cultures. Serious questions remain: If man did invent God for social reasons, how would those beliefs get into one's genes? While it is possible for one's genes to change behavior, there is no clear evidence to suggest the reverse—that one's behavior can change genes. When dentists pull wisdom teeth, they are not pulling genes, so people from the next generation will keep struggling with their wisdom teeth, for these "external" changes did not make it into the genes. Something similar might also hold for religion.

Reason #5. We are dealing here with genes that are allegedly connected with behavioral traits. Such hypothetical genes are supposed to control traits with very complex and variable patterns of behavior. They make for very intricate similarities that come in many varieties. Most of these hypothetical genes were once claimed, and then had to be retracted. So if we cannot even link individual genes to personalities, how can we possibly link genes to religion? There are too many intervening steps involved to make for such a simplistic link—for example, other genes,

12   *The Works of Voltaire: A Contemporary Version with Notes* (Charleston, SC: Nabu Press, 2010) volume 38, 15–19.

environmental effects on gene expression, cultural factors, upbringing, and personal experiences.

Even if we admit that genes do have an effect on how religious you are, upbringing still might have a big impact on the "brand" of religion you take on — not to mention experiences such as illnesses or the loss of dear ones. Just as genes don't tell you how to vote in elections, neither do genes tell you what to believe in religion. Curiously enough, Hamer's "god gene" accounted for only 1% of the variance in the test scores of his subjects, prompting Francis Collins, former head of the *Human Genome Project*, to say about Hamer's book that "a careful reading indicated that the title was wildly overstated."[13] Do we really believe we could compute a person's religion if we were given the entire DNA sequence and a large enough computer? That's hard to believe.

Reason #6. It is very doubtful whether all this genetic talk actually has anything to do with religion taken as a belief in a Transcendent Being, God. Perhaps genetics can tell us something about mystical experiences, but the idea that people believe in God because of mystical experiences is silly. One need not feel anything, let alone have a mystical experience, to believe in the existence of God. Arguably, most individuals who believe in God have never experienced God in a mystical way. Quite a few believe in God, or reject God, for purely intellectual reasons. Others simply have an intuitive awareness of God's existence. So the label "god gene" is very deceiving, to say the least — which Hamer did acknowledge himself, though. He admitted, "My findings are agnostic on the existence of God. If there's a God, there's a God. Just knowing what brain chemicals are involved in acknowledging that is not going to change the fact."[14] Kudos to Hamer for this confession.

Reason #7. A "feeling of transcendence" is not necessarily a religious experience, and if Hamer is right, it is in fact merely a biological one. The monoamines involved in the feeling of self-transcendence are the same monoamines that are jumbled by ecstasy, LSD, and other mind-altering drugs. If the feeling of transcendence is indeed a biological experience rather than a religious experience, then studies performed on that experience only tell us something about biology, not religion. The question

13 Francis Collins, *The Language of God* (New York: Free Press, 2006), 262.
14 Dean Hamer, *The God Gene*, 65.

of God's existence remains a philosophical and religious question, not a biological one. The core issue is whether religion comes from "above," or rather from "below" during the process of evolution. By reducing religion to spiritual feelings and emotions, to neuronal or neurological activities, Hamer is saying that religion is ultimately not what it appears to be.

While biology can tell us a lot about human beings, it does not and cannot tell us anything about God. Besides, spirituality may actually have little to do with God. It could very well be an emotional feeling, whereas religion is not just a feeling but rather a conviction about facts and truths as stated in a Creed, for instance. To put it more directly, God's existence is not dependent on our experience of him. The existence of God is a factual issue of yes or no. God either exists or he doesn't—that's not a matter of opinions or feelings or emotions. You can have your own opinions and feelings, but you can't have your own facts, not even in religion. Believing that the earth is flat does not make the earth flat. Believing that God does not exist doesn't make God disappear.

Reason #8. Religious belief is not merely a belief that is credible, believable, or acceptable to a certain degree, but it is the very basis on which our rationality and morality, all our thinking and doing, are ultimately based (see next section). Without this religious belief, rationality and morality would be baseless; we need some beliefs so that we may understand. Albert Einstein was right when he said that science without religion is lame (but he also rightly added that religion without science is blind).[15] It is faith in a Creator that makes us understand ourselves and makes us comprehend the Universe—which turns science into a faith-based enterprise, based on a belief in a Creator God who has given this world its order and its laws, so we can explore, understand, explain, and make predictions, because God created a "trustworthy" Universe. Therefore, denying that there is a Creator is actually an acid eating away not only the foundation of science, but of all rationality and all morality.

Our Universe must somehow possess an "inner logic" accessible to human reasoning—an intrinsic "rationality" that governs everything science is trying to decipher. Einstein expressed very emphatically his "confidence in the rational nature of reality and in its being accessible,

---

15   Albert Einstein, in *Science, Philosophy, and Religion, A Symposium* (New York: The Conference on Science and Religion, 1941).

to some degree, to human reason. When this feeling is missing, science degenerates into mindless empiricism."[16] At the very moment we claim that this belief is merely a product of genes we have undermined all our own activities, scientific and otherwise. Therefore, genetics as a science cannot explain religion or any religious beliefs, for it must assume certain religious beliefs before it can even start its explanatory job. It is thanks to God that scientists have reason to trust their own scientific reasoning. In light of this, one might argue that all scientists keep living off Judeo-Christian capital, whether they like to admit it or not. They borrow from the Judeo-Christian faith what they themselves cannot provide.

Reason #9. All of this prompts the question why Hamer wants to reduce religion and faith in God to something else, to something like spiritual experiences, and ultimately to genetic instructions. The answer can perhaps be found in some underlying philosophical assumption that he seems to entertain—the assumption that biology can fully explain everything in life, including religious faith, beliefs, and experiences. It is a deep-seated dream of certain scientists. The sociobiologist E. O. Wilson once triumphantly exclaimed, "[W]e have come to the crucial stage in the history of biology when religion itself is subject to the explanations of the natural sciences."[17]

This presupposition is based on the view that all reality can be reduced to a scientific explanation. However if there is only material stuff then there is not even the possibility of spirituality. Then there is no longer room for spiritual reality—perhaps for emotions and sensations that we consider spiritual, but not for a reality behind and beyond those experiences. Hence the "spiritual allele" of the "god gene" cannot be spiritual at all and can certainly not cause a religious experience. It may cause a natural sensation of self-transcendence, which some have unwittingly interpreted as an encounter with the Divine. If we had to give Hamer's gene a label, "hallucination gene" would probably be the best fit. Before Hamer can reduce spirituality to biology he must demonstrate that man is only flesh, rather than flesh and spirit, body and soul. Alas for him, science does not have that capacity. It may study man as if man is merely flesh, but it cannot transform man into mere flesh (see chapter 11).

16   Einstein's Letter to Solovine.
17   E. O. Wilson, *On Human Nature* (Cambridge, MA: Harvard University Press, 1978), 192.

Reason #10. Last but not least, we need to touch on a more general issue regarding any kind of research in genetics. Typically, geneticists assume that genes and their DNA determine many features and characteristics of an individual. That is surely a valid assumption, but it may easily become a dogma or creed, saying "It's all in the genes, and only in the genes!" James Watson, the co-discoverer of DNA, trumpeted this dogma to everyone: "We used to think that our fate was in our stars. Now we know, in large part, that our fate is in our genes."[18] At least he was honest enough to add the clause "in large part." Yet it should be stated that what this geneticist says about genetics is no longer genetics.

According to this ideology, even those kinds of behavior that we think are our own choosing—lifestyle choices, moral decisions, religious beliefs, and the like—would require us to postulate a gene for whatever we decide, believe, or choose. We labeled this before as the doctrine of genetic determinism (see chapter 8). Because it is rather rampant in genetics, new hypothetical genes just keep coming and going. We could even invent a gene that makes one believe in the all-powerfulness of genetics. That is where science borders science fiction, or actually turns into it. People who believe in strict genetic determinism—and a number of geneticists do, albeit a minority probably—are of the opinion that what some call their "free will" must, like anything else, be determined by genes and DNA as well. Why should we bother to claim anything at all, if all is predetermined anyway? As was said before, determinism is a doctrine that wants us to *believe* everything is predetermined. But how could it *make* us believe so, given the fact that our beliefs would be as predestined as anything else in life? Why bother to debate free will if people are already genetically determined to either believe in it or not?

Therefore, we can still maintain that we are not at the mercy of our genes. Our genes are not our destiny; they are like a hand of cards we are dealt, but we can play them differently. Human behavior is arguably more often than not a matter of lifestyle *choices* rather than the outcome of a set of genetic instructions. Humans are in control of those choices, at least to a certain degree. Human beings are free to choose whether to act or not to act—and in choosing, their choices *cause* them to act as much so as genes can cause us to act (but one is an immaterial cause,

---

18   Leon Jaroff, "The Gene Hunt," *Time Magazine,* March 20, 1989.

the other a material one). People who do not care about what is true or false and about what is right or wrong have *chosen* not to care; they have decided not to become a "good cause."

However, having a free will is essential for us to be masters of our actions, instead of victims — by allowing us to make our own choices and decisions in life, especially so when it comes to rationality, morality, and religion. To put it in more sneering terms, people who do not believe this must have a gene that makes them think that way.

RELIGION EXPLAINS EVERYTHING ELSE

Instead of exploring the genetic basis of religion, as we did in the previous section, it might be more promising to explore the religious basis of all we have discussed so far, including genetics. We take religion here as a belief in the existence of God — as a belief in the existence of a Primary Cause, of a Person above and beyond all there is. Of course, religion is more than just a belief like that, but it is also no less than that. Religion dramatically changes the way we see the world. As Ludwig Wittgenstein once said, "To believe in a God means to see that the facts of the world are not the end of the matter."[19]

We discussed already in this chapter that secondary causes — and genes are part of those — cannot even exist without a Primary Cause. The existence of God is the foundation and explanation of everything there is in this Universe. Without God, there is nothing, not even genes. Nothing can exist on its own. Nothing can pop up on its own. The contrary idea that something can cause itself and explain itself has rightly been caricaturized by the Boston College philosopher Peter Kreeft as a magical "pop theory" which has things pop into existence without any cause.[20] Nothing — not even a Higgs boson (once called "the God particle") — can just pop itself into existence; it must have a cause, because it does not and cannot have the power to make itself exist. For something to cause itself to exist, it would have to exist before it came into existence — which is logically and philosophically impossible. Nothing can just pop up out of nothing — nothing comes from nothing, as the saying goes.

19 Ludwig Wittgenstein, *Notebooks 1914–1916*, journal entry July 8, 1916, 74e.
20 Peter Kreeft, *The Creed: Fundamentals of Christian Belief* (San Francisco: Ignatius Press, 1988), 27.

Recently, scientists working at CERN declared that the universe as we know it should not exist. Ironically, they could have known this without any scientific research. If we deny the existence of God, the universe indeed could not even exist. We have no scientific reason why the universe does exist—and yet it does exist. God is a First, Primary, or Transcendent Cause outside and above the sequence of secondary causes. Without God, nothing in this Universe could exist. Even if we were able to explain each secondary cause by another cause, we still have the unanswered question left of what explains this very sequence of causes. Without a First Cause, none of the secondary causes could exist, let alone become causes themselves.

In this section, we want to get more detailed and specific on this issue. First, we will demonstrate that our faculty of rationality (as discussed in chapter 7) would be baseless without God's existence. Then, we will show something similar for our faculty of morality (as discussed in chapter 8). The thesis here is that both faculties require the existence of God for their foundation and explanation. Without God they would collapse.

Let's start with rationality. How is it possible that our faculty of rationality allows us to understand the world around us the way it is? Is there some mysterious match between the way we think about the world and the way the world actually is? More particularly, how is it possible, for instance, that our way of reasoning in terms of "like causes have like effects" corresponds to the causality we find operational in this Universe?

To assume that this is only a matter of thinking *habits* we have misses the point entirely. David Hume thought that the "mechanism" of causality is merely a kind of illusion produced by habit or custom. We simply imagine causality—in his own words, "From causes which appear similar we expect similar effects."[21] If that were true—that is, if effects were just our expectations—then causality would be the same as correlation—which is unacceptable in science. If Hume's idea were true—that is, if laws of nature were just mental habits—then it would be hard, if not impossible, to explain why bridges built in accordance with the proper physical laws stand firm whereas others collapse. It is hard to believe that competent engineers merely have better mental habits than their inept colleagues.

---

21 David Hume, *An Enquiry Concerning Human Understanding* (Indianapolis, IN: Hackett Publishing Co., 1993 [1772]), section V, part 1.

The laws of nature that scientists have come up with must be more than fanciful creations of the mind. We need bridges built by engineers with the right knowledge, not with better habits or opinions.

So the question is: Where do the laws of nature come from, if they are not just creations of the mind? How is our faculty of rationality — working with reason and intellect — capable of grasping reality as it is? Somehow the mind and the Universe were made for one another. Mind is a faculty that has the potential for understanding the Universe, and the Universe is an entity that has the potential for being understood by mind. The order found in the Universe — based on causality and laws of nature — matches the order found through our rationality — based on intellect and reason. Obviously there is order in the Universe, otherwise it would make no sense to search for laws of nature. Only if there is "law and order" in nature are scientists able to explain and predict. Because we "know" through reason that like causes produce like effects, our faculty of rationality is able to create explanations and predictions.

The order found in the Universe — based on causality and laws of nature — is not something science has discovered after a long, extensive search. It is rather something we *assume* before science can come off the ground. As Dimitri Mendeleev, who discovered the Periodic Table of Elements, put it: "It is the function of science to discover the existence of a general reign of order in nature and to find the causes governing this order."[22] Science can never prove there is order in this Universe, but instead must adopt order as a universal principle before it can prove anything else. Even the rule of falsification in science is based on order and cannot be falsified by disorder, otherwise counter-evidence would not allow us to falsify theories. Therefore, counter-evidence can falsify a hypothesis, but never the principle of falsification itself.

The assumption of order in nature is a "prerequisite" for scientific knowledge, actually for any kind of knowledge. It is not an *a priori* form of thought but a *given* rooted in reality, "engraved" in the structure of the Universe. It is a given that enables intelligibility, thus allowing us to grasp reality by using reason and intellect. Even when we haven't come up with an explanation of why something happened, we don't doubt that

22 Dmitri Mendeleev, *Mendeleev on the Periodic Law: Selected Writings, 1869–1905* (Mineola, NY: Dover Publications, 2005).

there is one, and we often do have an explanation for why we don't have an explanation. But does *everything* have an explanation, even if it's an explanation we haven't discovered yet and perhaps never will discover? Yes, we know there is an explanation for everything — not just because we have found it to be so over and over again, but because we *assume* that's so — similarly to the way we assume there is order in this Universe. If there were no order and no possibility of explanation, we couldn't even trust and explain our own perceptions and observations.

This presumed order is actually rather astonishing. Albert Einstein wrote in one of his letters: "But surely, *a priori*, one should expect the world to be chaotic, not to be grasped by thought in any way."[23] However, if the world were not orderly or intelligible, we couldn't even trust the evidence of sensory perception, nor the empirical sciences grounded in perception. To reject that the world is orderly and intelligible undermines the possibility of any rational inquiry. It would be the end of any empirical science grounded in perception and observation.

And yet thought and rationality, reason and intellect, are able to grasp the order of reality and explain it. Where does this harmony between rationality and reality come from? Or put more specifically, how can it be that mathematics, being after all a product of human thought independent of experience, is so admirably appropriate to the objects of reality around us? Put differently, the physical order we observe in this world appears to be amazingly "consistent." Not only is our rationality consistent, but so is the world itself. It is a consistency that must perplex us, for how is it possible that reality can be "grasped" by the rationality of the human mind? We can just ignore this question, of course, but that would be irrational. There is a certain step of faith required in putting all one's intellectual weight on the pedestal of reason. The most fundamental assumption of the rational mind is that the world perceived through reason is true — that the world itself is "reasonable." This presupposes that the mind interpreting the world through reason is somehow apprehending the world as it actually exists.

The most likely explanation of this harmony is that there is not only rationality in the human mind but also "rationality" in the world itself. This gave the late astrophysicist Sir James Jeans cause to say: "[T]he

---

23 Albert Einstein, Letter to M. Solovine.

Universe begins to look more like a great thought than a great machine."[24]
Even the physicist Paul Davies dared to say at one point: "There must
be an unchanging rational ground in which the logical, orderly nature
of the Universe is rooted."[25] Even to say that the world is ultimately
governed by some fundamental laws of nature does not explain anything, but would just boil down to an unintelligible "brute fact." The laws
of nature could have been other than they are, so they cannot explain
themselves. Not explaining where the laws of nature come from—but
treating them as "brute facts" instead—makes everything else based on
those laws ultimately unexplained and unintelligible. The existence of
anything, including laws of nature, must have an explanation. The laws
of nature cannot explain the existence of the Universe for the simple
reason that they *presuppose* the existence of the Universe.

In other words, there seems to be some mysterious conformity here
between the rationality of our minds and the "rationality" found in the
world around us. Somehow the mind seems to be able to capture nature
the way it *is*. This leads us inevitably to the question: Where does the
"rationality" of the world come from? The only rational explanation seems
to be that there is indeed an intelligent, rational, orderly, and lawgiving
Creator God who made this Universe the way it is. How could nature be
intelligible if it were not created by an intelligent Creator? How could
there be order in this world if there were no orderly Creator? How could
there be scientific laws if there were no rational Lawgiver?

So we end up with a rather perplexing conclusion: We actually need
faith in a Creator God to explain why our faculty of rationality actually
works. This faith in God and his existence can explain why there is an
almost perfect harmony of thought and being, of rationality and reality.
One could even make the case that denying or neglecting the existence of
God would undermine rationality and thus eat away the very foundation
of all we think we know, including all we know through science. Without
God, we would have no reason to trust our own reasoning.

God is the "author" of reason. Belief in God and in God's existence
makes the world so much more understandable and intelligible for the

24 James Jeans, *The Mysterious Universe* (Cambridge: Cambridge University Press,
1930), chapter 5.
25 P. Davies, "What happened before the big bang," in Stannard, R. (ed.), *God for
the 21st Century* (West Conshohocken, PA: Templeton Press, 2000), 10–12.

human intellect. The astronomer Johannes Kepler worded this beautifully: "The chief aim of all investigations of the external world should be to discover the rational order and harmony which has been imposed on it by God."[26] Even Albert Einstein had to acknowledge: "Everyone who is seriously involved in the pursuit of science becomes convinced that a Spirit is manifest in the laws of the Universe — a Spirit vastly superior to that of man."[27]

So we must come to the conclusion that the existence of God — rather than the theory of genetics or the theory of natural selection — is the foundation and explanation of the human faculty of rationality. When instead we downgrade religion to the status of being the mere product of genetics and natural selection, we forget that the very theory of natural selection must then also be the mere product of natural selection, which makes it a doctrine that ultimately devours itself. It is hard to imagine how natural selection could ever produce the *theory* of natural selection. That would be like the miracle of a hand drawing itself; a hand may draw a picture of a hand, of course, but a hand that draws a hand cannot produce the very hand that does the drawing. The very hand that does the drawing must be "more" than the hand that is being drawn on paper. Likewise, the intellect that came up with the theory of natural selection must be more than a product of natural selection.

Even Darwin himself was aware of this thorny problem. Here is what he said in his *Autobiography*: If the theory of natural selection comes from the human mind, one might wonder whether "the mind of man, which has, as I fully believe, been developed from a mind as low as that possessed by the lowest animal, [can] be trusted when it draws such grand conclusions."[28] The obvious answer to this rhetorical question seems to be No. And again in an 1881 letter: "Would anyone trust in the convictions of a monkey's mind, if there are any convictions in such

26　*De fundamentis Astrologiae Certioribus*, Thesis XX. Also quoted in Morris Kline, *Mathematical Thought from Ancient to Modern Times* (New York: Oxford University Press, 1990), 231.

27　*Albert Einstein's Letters to and from Children*, ed. Alice Calaprice (Amherst, NY: Prometheus Books, 2002), 127–29. Also quoted in: Max Jammer, *Einstein and Religion: Physics and Theology* (Princeton: Princeton University Press, 1999), 93.

28　Charles Darwin, *The Autobiography of Charles Darwin* (Cambridge, UK: Icon Books, 2003), 149.

a mind?"[29] The obvious answer to this rhetorical question would also be "No one would." (Let's forget about his deceiving use of the word "mind." Chapter 11 has more on this.)

Curiously enough, in these passages Darwin applied this insight to one's belief in God, but not to his own belief in evolution and natural selection. Somehow his own theory ended up on his "blind spot." Therefore, Darwin concluded that we cannot trust anything we claim to know about God. However, one could as well, or even better, argue the larger claim — that we could not trust anything we know at all if there were no God. If natural selection, instead of God, were the origin of all there is, including the human mind, it would act as a boomerang that comes back to its maker in a vicious circle, knocking out the truth claims of whoever launched them. How could we ever trust the outcome of mere natural selection when it comes to matters of truth? In fact the theory of natural selection must *assume* the human mind with its faculty of rationality, but it can neither create it nor fully explain it. On its own, natural selection would be just a powerless and useless concept, for if one can't trust the rationality of human beings, one is logically prevented from having confidence in one's own rational activities — with science being one of them, including the theory of natural selection.

It is time now to proceed to our second issue: the faculty of morality (as discussed in chapter 8). We will demonstrate here that this faculty would also be baseless without God's existence; it too requires the existence of God for its foundation and explanation.

If our human faculty of morality could be completely reduced to genes, or any other natural factors, we would have to face some unacceptable thorny consequences. How would we then explain why morality is such a demanding issue — indeed demanding an absolute authority? This would be impossible if morality were only a matter of genes, brain wiring, tradition, majority votes, or political correctness. Do our genes, or any other natural factors, have the right to demand absolute obedience from us? Of course not! Does society or the government have the right to demand our absolute obedience in matters of moral rights and duties? Certainly not! Does any person have the right to demand our absolute obedience? None of the above!

29   Darwin's letter to W. Graham, 1881.

Yet morality comes with absolute demands. Even moral relativists, who deny that morality has any absolute authority, still hold on to at least one *absolute* moral demand: "Never disobey your own conscience." Almost all people have something about conscience that they respect, even if their theory is that it's nothing. Other moral relativists have replaced the moral code of the Ten Commandments with what sociologist Alan Wolfe calls a new, 11th commandment, "Thou shalt not judge."[30] Somehow, even moral relativists come up with moral statements that everyone is supposed to accept as universal and absolute. There must be some absolute and universal moral demands that transcend race, ethnicity, nationality, culture, religion, and political affiliation. If that were not the case, for example, there would not have been any justification for the Nuremberg trials that took place after World War II — or for any other international court, for that matter.

In other words, we need a foundation and explanation of morality that transcends everything as relative as genes, brain wiring, tradition, majority votes, or political correctness. Where can that foundation be found? There seems to be only one "place" where this authority can be found: in Heaven. The best, and arguably only, rational explanation would be that morality does not come from "below" — let alone from the animal world — but must come from "above." The only authority that can obligate us is someone infinitely superior to us; no one else has the right to demand our absolute obedience. In other words, morality comes from God and his Natural Law; and therefore, there can be no morality without God and his Natural Law. Ironically, for relativists, the Natural Law with its objective, absolute, and universal demands (see chapter 8) is the dreaded "N-word."

Perhaps surprisingly, even an atheist such as the late French philosopher Jean-Paul Sartre realized that there can be no absolute and objective standards of right and wrong, if there is no eternal Heaven that would make moral laws and values objective and universal. As an atheist he had to conclude, though, that it is "extremely embarrassing that God does not exist, for there disappears with him all possibility of finding values in an intelligible heaven. There can no longer be any good *a priori*, since

---

30   Alan Wolfe, *One Nation, After All* (London: Penguin Books, 1999).

there is no infinite and perfect consciousness to think it."[31] Because Sartre denied the existence of God for most of his life, he realized very clearly that by being an atheist he also had to give up on morality. If there is no God, there cannot be evil either. As Thomas Aquinas famously said, "Good can exist without evil, whereas evil cannot exist without good."[32]

The German philosopher Friedrich Nietzsche was another atheist who realized how devastating the decline of religion is to the morality of society when he wrote: "God is dead; but as the human race is constituted, there will perhaps be caves for millenniums yet, in which people will show his shadow."[33] Nietzsche is saying here that humanism and other "moral" ideologies shelter themselves in caves and venerate shadows of the God they once believed in; they are still holding on to something they cannot provide themselves — mere shadows of the past. These are "idols" constructed to preserve the appearance of morality without the substance.

Nietzsche clearly understood that "the death of God" meant the destruction of all meaning and value in life. Once we think we can understand the world apart from God, God is dead in the way Latin is dead. Nietzsche saw in all clarity how in a world without divine and eternal laws neither our dignity nor our morality would in the long run be able to survive. In contrast, Jürgen Habermas, although a non-religious philosopher, expressed as his conviction that the ideas of freedom and social co-existence are based on the Jewish notion of justice and on the Christian ethics of love. As he puts it: "Up to this very day there is no alternative to it."[34] This does not mean, of course, that we must believe in God in order to live a moral life. As Nietzsche put it, we can still venerate "idols from the past." But without God, moral life has lost its foundation.

Because of all of this we must recognize that morality can ultimately come only from "Above." Moral laws and values reside in Heaven. That's where their universality and objectivity come from. We ought to do what we ought to do — for Heaven's sake! The *United States Declaration of Independence* is in tune with this when it declares that we are endowed by our

---

31   Jean-Paul Sartre, *Existentialism from Dostoevsky to Sartre*, ed. Walter Kaufmann (New York: New American Library, 1975), 353.

32   *Summa Theologiae* I, 109.1 ad 1.

33   Friedrich Nietzsche, *The Gay Science*, transl. Walter Kaufmann (Vancouver, WA: Vintage Books, 1974), §108 and §125.

34   Jürgen Habermas, Ciaran Cronin, and Max Pensky, *Time of Transitions* (Cambridge: Polity Press, 2006), 150–51.

Creator with certain unalienable Rights—not man-made but God-given rights, that is. When in 1948 the United Nations (UN) affirmed in its *Universal Declaration of Human Rights* that "all human beings are born free and equal in dignity and rights," it must have assumed the same without explicitly mentioning it (the drafters famously left the term "right" vague in order to achieve passage). Without this assumption all those rights would be sitting on quicksand, subject to the mercy of lawmakers and majority votes. But the Catholic philosopher Jacques Maritain, a Thomist actively involved in drafting the *United Nations Universal Declaration of Human Rights*, was right when he said paradoxically: "We agree on these rights, on condition that no one asks us why."[35] When we do ask why we have human rights, the only reasonable answer is: because God has endowed us with rights.

It is through the voice of God, in the *Natural Law*, that we know about right and wrong, about rights and duties (see chapter 8). Without God, who is the author of human rights, we would have no right to claim any rights. If there were no God we could not defend any of those rights we think we have the right to defend. Instead we would only have (legal) *entitlements*, or privileges, which the government provides us with, but no (moral) *rights*, which only God can provide. Here we have the stark difference between the universal Natural Law, on the one hand, and the various local laws (legal, civil, or positive laws) on the other hand. John F. Kennedy put it well in his Inaugural Address: "the rights of man come not from the generosity of the state, but from the hand of God." Without an eternal Heaven there could be no absolute or objective standards of right and wrong. If these did not come from God, people could take them away anytime—which they certainly have tried many times to do.

Let's come to a conclusion about the foundation and explanation of rationality and morality. Only the existence of God can explain that there is a Universe, that there is order in this Universe, that this Universe is intelligible, that there are universal laws of nature, and that there are universal moral laws. This is the only way we can take the world as something created according to an intelligible plan accessible to the human intellect through the natural light of reason. This is the only way we

---

35 Jacques Maritain, *Man and the State* (Chicago: University of Chicago Press, 1951), 90–100.

can explain the natural order and moral order of this Universe. Because there is a Creator, we have not only a rational Lawgiver—who guarantees order, intelligibility, and predictability—but also a moral Lawgiver—who guarantees decency, integrity, conscience, responsibility, justice, and the human dignity that comes with human rights. C.S. Lewis beautifully summarized this: "I believe in Christianity as I believe that the sun has risen: not only because I see it, but because by it I see everything else."[36]

Therefore, denying that there is a Creator God is actually an acid eating away not only the foundation of science, but the foundation of rationality and morality also. Our Universe possesses an "inner logic"—a natural order and a moral order—accessible to human intellectual and moral reasoning. At the very moment we claim that this belief is merely a product of genes, we have undermined all our own activities, scientific and otherwise. Without God, scientists would fundamentally lose their *reason* for trusting their own scientific reasoning. Without the notion of the Universe as a created entity, science would become a shaky and problematic enterprise. Without the notion of a Heaven with absolute and universal moral values and laws, everything would be permissible, and we would have no *right* to claim any moral rights. In short, if we lose the notion of Creation, plus the trustworthy order it comes with, natural laws and moral laws would become utterly questionable and problematic.

In other words, religious belief in God's existence is not merely a belief that is credible, believable, or acceptable to a certain degree, but is the very basis on which our rationality and morality, all our thinking and doing, are grounded. Without this religious belief, rationality and morality would be baseless and collapse. Is that a *proof* of God's existence? In a way, it is!

The best we can say right now is that there is a deep religion divide between the pre-human and human world. There is no way of telling whether this leap of religion can be fully explained in terms of genes and mutations. Obviously, the faculty of religion requires a body with the "right" features. The "right" mutations were perhaps *necessary* to achieve this. But whether those mutations were also *sufficient* is impossible to tell ahead of time, but most unlikely. Those who think the religion divide can

---

36 C.S. Lewis, "They Asked For A Paper," in *Is Theology Poetry?* (London: Geoffrey Bless, 1962), 165.

eventually be explained completely by science must realize this cannot be more than a conviction — more of a program than an achievement. There is no scientific way of corroborating such a conviction — it is worth as much as the opposite claim that a scientific explanation for our human faculty of religion will never be fully possible. But what remains standing is the fact that without religion we have no reason to trust any scientific reasoning whatsoever.

11

The Soul Divide

THE PREVIOUS CHAPTERS SUGGEST THAT HUMANS differ from non-human animals by language, rationality, morality, self-expression, and religion. If it is true that these distinctive human features did not come from their pre-human ancestors or from their pre-human genes, then the question arises: Where could these distinctive human features come from? The Catholic answer is: from their immortal souls— and their immortal souls come directly from God.

MORE THAN MATTER

How can there be a soul in a world of material things? In a world of pure matter, there seems to be no room for souls. Has science not shown us that everything in this Universe can be quantified, measured, counted, and dissected? If so, then anything else cannot be real, or at least should not be worth our attention — including souls.

The incentive behind these questions is some kind of worldview that is usually called *materialism*: the philosophical idea, or even doctrine or dogma, that the only thing that exists is *matter*; that all things are composed of matter, and that everything is the mere result of material interactions. Its slogan is: Everything that exists is matter, and matter is all there is. If this were true, then that would be the end of anything non-material.

This immediately poses the next question: What is this "matter"? Matter is anything that has mass and takes up space. Although "matter" is a very old concept, it has lost its prominent position in physics, because the term "mass" is well-defined, but "matter" is not. Yet the strength of materialism is that it centers on one of the most noticeable elements in the world around us — "stuff," that is. It is that which can usually be seen, touched, heard, tasted, and smelled by our five senses. It is everything that can be quantified, measured, counted, and dissected, particularly in science. Studying this kind of "stuff" has been vital for our survival. It is impossible to deny that almost all our technological achievements are based on it. We cannot disregard it, we cannot live without it. Where would we be without this "stuff"?

That's basically the reason why materialism has become so popular. In this view we are only able to know and study material objects that can be perceived by the five senses. Because we cannot know non-material things — so do believers in materialism say — we must conclude that non-material things do not and cannot exist. Their creed is "Everything that exists is material," for nothing can exist without the materials out of which it is made. The material world is, in the words of Carl Sagan, "all that is, or ever was, or ever will be."[1] Therefore, what materialism doesn't detect does not and cannot exist.

Indeed, materialism seems to be an unbeatable idea. The world of our sense-experience is filled with nothing but individual objects, all of which are physical bodies, material things, or their attributes. The fact that all the things we perceive through our senses are individual physical things or material embodiments gives great credibility to the materialistic doctrine that the world of real existences and entities is entirely material, and that nothing immaterial really exists or that it ends up being merely fantasy and fiction.

Yet on further inspection there are several serious and potentially detrimental problems with materialism.

First of all, to think that materialism is backed by science is a misconception. The most science can give us is the observation that *many* things in this Universe are material and can be quantified, measured, counted, and dissected. However, the step from "*many* things" to "*all* things" is infinitely large — there is no way we could conclusively reason from "many" to "all." Many swans are white, but it is impossible to prove that all swans are white. So the conclusion that *all* things in this Universe are material is not logically justified; it is at best of an inductive nature — but certainly not of a logically safe deductive nature. There is no way of knowing. It takes us on a never-ending search for proof.

Second, materialism comes close to circular reasoning. Science on its own can never prove that matter is all there is, because science first limits itself exclusively to material things, and then says science shows us there *is* actually nothing but matter. It excludes immaterial entities already prior to our discourse, and then "concludes" there are only material

1   With these words Carl Sagan opened his series Cosmos in 1980. New book edition: Carl Sagan, Ann Druyan, and Neil deGrasse Tyson, *Cosmos* (New York: Ballantine Books, 2013), 4.

entities. That is an example of circular reasoning: It begins with what it is trying to end with — that's how we keep circling around. At best, materialism is a metaphysical assumption — and a bad one at that, as we will explain further.

Third, materialism is monopolistic. It is a totalitarian ideology, for it allows no room for anything but itself. It asserts not only that matter is everywhere, but also that matter is the only thing there is. It allows for only one way of looking at the world — one exclusive and comprehensive way. Thus it suffers from megalomania. It is monopolistic by nature, and doesn't tolerate competitors. It is not just *a* method — one out of several — but also a claim that this is the only valid method. However, no matter how well materialistic models explain, for instance, the workings of a human body, that doesn't mean a human being *is* actually a machine. Looking at it as if it were a machine — which may be a very successful approach for scientists — does not *make* it a machine. There should always be room for other views and perspectives on the world. Nothing entitles us to deny that there is more to life than the material entities of molecules, neurons, and genes.

Fourth, materialism is inconsistent. It expresses a *thought* some people have — a thought with no mass, no size, no color. Suddenly, something non-material pops up in the materialistic world of materialism — something that is not small or heavy but true or false. If there were no world of true and false, good logic would be as misleading as bad logic. If materialists want to claim materialism is true, they should realize that, for it to be true, this very thought must be more than a certain pattern of electrical activity in their brain cells. People have known the contents of their own minds from time immemorial without knowing anything about brains or genes. So thoughts must be more than material brain waves and genetic DNA code — in the same way as love must be more than a chemical reaction. The burden of proof is on whoever denies this. The famous philosopher Thomas Nagel came to the daring conclusion that the materialist conception of nature is almost certainly false, except for those who deny that the mental is an irreducible aspect of reality.[2] But the question is, what entitles us to declare the mental identical to the neural (see chapter 7)?

2   Thomas Nagel, *Mind and Cosmos: Why the Materialist Neo-Darwinian Conception of Nature Is Almost Certainly False* (Oxford: Oxford University Press, 2012).

Fifth, materialism is self-destructive. The statement that there are only material things is as fragile as is the "material" that supposedly generated this worldview. It declares that things can only be explained in terms of material entities such as molecules, genes, and neurons. However, as John C. Polkinghorne puts it, "neural events simply happen, and that is that."[3] A pattern of electrical nerve impulses can't be true or false. As Alvin Plantinga puts it, "It's a little like trying to understand what it would be for the number seven, e.g., to weigh five pounds.... A number just isn't the sort of thing that can have weight."[4] We can think about the size, weight, or color of things, but these thoughts themselves do not have a size, weight, or color. Thoughts are immaterial — not something the brain secretes. If the mental were the same as the neural, thoughts could never be right or wrong and true or false. Reducing concepts and thoughts to a "creation of neurons in the brain" obscures the fact that "neuron" is itself an abstract, immaterial concept. Such a claim starts a vicious circle. Stephen Barr shows us the vicious circle: "The very theory which says that theories are neurons firing is itself naught but neurons firing."[5]

Sixth, materialism is insufficient. Francis Crick may be correct to exclaim that "without understanding molecules we can only have a very sketchy understanding of life itself," but materialism makes the logical mistake of reversing this statement: Understanding molecules is supposed to give us a complete understanding of life. That is not logically warranted. Understanding life is more than understanding molecules, for there is arguably more to life than molecules. In addition, there is more to human life than animal life. What goes beyond molecules, for example, are the concepts and beliefs human beings foster, no matter whether it is in science or in religion. So it is okay — in fact methodologically mandatory — for scientists to deal only with the material aspects of this Universe, but this does not entitle them to claim there is nothing more in this Universe. A sign hung in Albert Einstein's office at Princeton University advised us to rise above materialism: "Not everything that can be counted counts; not everything that counts can be counted."

3 John Polkinghorne, *Science and Creation — The Search for Understanding* (Boston: Shambhala, 1989), 36.
4 Alvin Plantinga, "Against Materialism," *Faith and Philosophy*, vol. 23, (Jan. 2006): 14.
5 Stephen Barr, *Modern Physics and Ancient Faith*, 196.

Seventh, materialism leads to contradictions. The biologist J.B.S. Haldane described the following contradiction: "If materialism is true, it seems to me that we cannot know that it is true. If my opinions are the result of chemical processes going on in my brain, they are determined by chemistry, not the laws of logic."[6] Materialism is merely a philosophical or metaphysical position; it cannot be a conclusion of the empirical sciences. Those who deny the existence of anything immaterial thereby also deny the existence of their very own denial, since all statements, including denials, are immaterial—they are either true or false, as we said earlier. C.S. Lewis exposed the circularity of this claim as follows:

> A theory which explained everything else in the whole Universe but which made it impossible to believe that our thinking was valid, would be utterly out of court. For that theory would itself have been reached by thinking, and if thinking is not valid that theory would, of course, be itself demolished. It would have destroyed its own credentials. It would be an argument which proved that no argument was sound—a proof that there are no such things as proofs—which is nonsense.[7]

Eighth, materialism is meaning-less. This may sound strange, at first sight, but earlier (in chapter 7) we discussed that concepts always have meaning (and sometimes reference). Well, materialism is the end of that; it discards meaning. Meanings are ideas that are understood by minds. However, if the word "snow" were merely something in our head, then how could we ever communicate with others regarding snow? We would never mean the same thing whenever we talk about snow—we couldn't mean what we say nor say what we mean. So when materialism reduces mental ideas to mere patterns of electrical impulses in the brain, it has put an end to meaning and sense. That's why we can call materialism meaning-less, sense-less, and nonsensical. If there are no longer concepts with meanings—but only neurons firing or genes instructing—then all of scientific thought would be reduced to neurons firing or genes instructing. G.K. Chesterton rightly called this "the suicide of thought."[8]

6   J.B.S. Haldane, *The Inequality of Man*, 162.
7   C.S. Lewis, *Miracles* (London: Fontana, 1963), 18.
8   G.K. Chesterton, *Orthodoxy*, chapter 3.

Ninth, matter as conceived by materialism is not self-explanatory the way materialists think it is. Material things cannot account for their own existence for the simple reason that matter is that which by definition already exists. Yet, materialists seem to claim that matter is self-explanatory. They may not explicitly express it that way, but for them matter is a "primary cause" that needs no further explanation, and is responsible not only for things coming into existence but for their continuing in existence as well, because nothing can exist, so they say, without the materials out of which it is made. So they replace God as a primary cause with matter.

What is wrong with that position? Since matter is anything that has mass and takes up space, it is subject to motion and change. It is contingent, not necessary. If it were necessary, then everything about matter could be deduced by pure thought without doing any observations or experiments — which is absurd. The Primary Cause, on the other hand, cannot be subject to motion and change. Michael Augros brings this argument to a close: "Matter itself is a *product*, receiving its very existence from the action of something before it."[9] So God as the only Primary Cause, uncaused and motionless, remains standing; it is God who brings matter into existence and keeps it in existence. Matter cannot do so on its own. That's why matter cannot be a primary cause. Nothing, not even matter, can just pop itself into existence; it must have a cause, because it does not and cannot have the power to make itself exist. As said earlier, for something to cause itself to exist, it would have to exist before it came into existence — which is logically and philosophically impossible. Therefore, matter is not self-explanatory and thus cannot be a Primary Cause.

Let's come to a more general conclusion. Matter may be everywhere, but it is certainly not all there is. If matter were indeed all there is, then one should wonder what materialism itself is. Another piece of matter? If not, there must be more than matter. This definitely leaves room for non-material things such as logic, mathematics, philosophy, rationality, morality, and ultimately religious faith. There is so much more in life that the thermometers, Geiger counters, and DNA-probes of materialism can never capture — things such as souls, thoughts, concepts, values,

---

9 Michael Augros, *Who Designed the Designer?*, 63.

beliefs, laws, hopes, dreams, and ideals. There is no way materialism can deal with these—other than denying them, but then it must deny itself as well. God is "the Creator of Heaven and Earth, of all that is visible and invisible," in the words of the Creed—that is, not only of all that is material but also of all that is immaterial, which includes the soul.

What we have seen so far suggests that the faculties of language, rationality, morality, self-awareness, and religion may not come with matter—genes, neurons, hormones, enzymes—but may be part of the immortal soul. When did such souls emerge in evolution or prehistory? The answer to this question is far from obvious—for the simple reason that, as astronomer Owen Gingerich pithily puts it, "the transition to a spiritual being...does not fossilize."[10]

THE POWER OF THE MIND

The soul is one of those entities that materialism denies, but its existence can hardly be denied when it comes to explaining the faculties of language, rationality, morality, self-awareness, and religion. These faculties are all tightly linked to what we call the human *mind*. The human mind is the intellectual part of the human soul. If the mind does not exist, all the uniquely human faculties mentioned so far may end up being a mere illusion. But on the other hand if the human mind does exist as the intellectual part of the human soul, and thus makes for an undeniable reality, then all these faculties find their origin in the human mind (and ultimately in the human soul). So the discussion as to whether mutations can explain the origin of these faculties ultimately depends on how "real" the human mind is.

It is needless to say that mutations can change the *brain*, but the question remains whether they can also change the *mind*. Obviously, the Neo-Darwinian view is that mutations have indeed changed the brain and that all our "unique" faculties can be traced back to mutations that altered the brain. In other words, if the mind is identical to the brain, and if mental issues are nothing but neural issues, then the discussion is closed and everything that happened to the first humans can be explained by genetics. But if the mind is not the same as the brain, the discussion

10 Owen Gingerich, *God's Planet* (Cambridge, MA: Harvard University Press, 2014), 91.

is far from over. We have reached here a critical point in this debate. The pivotal question is this: Is the mind identical to the brain? Many scientists, especially those of the Neo-Darwinian paradigm, would say it is — they are basically "brain-mind equalizers." But what would happen if they are wrong?

Well, there are several reasons why it may not be possible to equate or reduce *mental* phenomena to *neural* phenomena. Some of these reasons are scientific, some empirical, some epistemological, and some just common sense. To find out what they are worth, let's study them in more detail.

Reason #1. One of the pioneers in neurosurgery, Wilder Penfield, made a compelling case about the difference between mental events and neural events, when he asked one of his patients during open-brain surgery to try and resist the movement of his left arm, which Penfield was about to make move by stimulating the motor cortex in the right hemisphere of the patient's brain. Thereupon the patient grabbed his left arm with his right hand, attempting to restrict the movement that was to be induced by a surgical stimulation of the right brain. As Penfield described this, "Behind the brain action of one hemisphere was the patient's mind. Behind the action of the other hemisphere was the electrode."[11] In other words, one action had a physical, neural cause whereas the other action had a mental cause. Therefore, he concluded, the physical cause and the mental cause had a different origin and were of a different nature.

From this follows, as neurologist Viktor Frankl put it, that while the brain does condition the mind, it does not give rise to it. The neurophysiologist Sir John Eccles concluded from experiments like Penfield's, "voluntary movements can be freely initiated independently of any determining influences within the neuronal machinery of the brain itself."[12] This observation seems to call for a mind in addition to the brain. The cognitive scientist Jerry Fodor put it most vividly and dramatically: "If it isn't literally true that my wanting is causally responsible for my reaching, and my itching is causally responsible for my scratching, and my believing

---

11   Wilder Penfield, in the *Control of the Mind* Symposium, held at the University of California Medical Center, San Francisco, 1961. Quoted in Arthur Koestler, *Ghost in the Machine* (London: Hutchison Publishing Group, 1967), 203–4.

12   Sir John Eccles and Karl R. Popper, *The Self and Its Brain* (New York: Springer Verlag, International, 1977), 294.

is causally responsible for my saying ... if none of that is literally true, then practically everything I believe about anything is false and it's the end of the world."[13]

Reason #2. When neuroscientists claim that certain mental phenomena are associated with certain neural phenomena, they cannot conclude from this that these mental phenomena were *caused* by neural phenomena. The reason is that correlation doesn't automatically equal causation. The use of umbrellas, for instance, is correlated with rain, but umbrellas don't cause rain. In a similar way, the fact that regions light up during functional magnetic resonance imaging (fMRI measures brain activity by detecting associated changes in blood flow) does not explain whether this lit-up state indicates they are causing a certain mental state, or just reflecting it.

Nevertheless, brain-mind equalizers use these so-called localization studies to make their case. When certain kinds of mental activity occur, certain parts of the brain do display increased blood flow and increased electrical activity. This makes them conclude that it is the brain that does the thinking, so thinking is merely a neural, not mental, activity. However, Alvin Plantinga, for one, does not buy this conclusion: "There are many activities that stand in that same or similar relation to the brain. Consider walking, or running, or ... moving your fingers: for each of these activities too there is a part of your brain related to it in such a way that when you engage in that activity there is increased blood flow in that part.... Who would conclude that your fingers' moving is really an activity of your brain and not of your fingers? Your fingers' moving is dependent on appropriate brain activity; it hardly follows that their moving is just an activity of your brain."[14] Then he comes to the conclusion that the mind's *dependence* on the brain is one thing, but *identity* between mind and brain quite another.

As we said earlier, it is true that we cannot understand without using our brains, but it does not follow that our brains are doing the understanding. In other words, brain activity may be a necessary condition for mental activity, but it does not seem to be a sufficient condition. In fact there are situations where the most intense subjective experiences

13   Jerry Fodor, *A Theory of Content and Other Essays* (Cambridge, MA: Bradford Book/MIT Press, 1990), 156.

14   Alvin Plantinga, "Against Materialism," *Faith and Philosophy*, vol. 23, (Jan. 2006): 23.

correlate with a dampening—or even cessation—of brain activity. What comes to mind are cases of Near-Death Experiences (NDEs) or Out-of-Body Experiences (OBEs) induced by G-LOC (or G-force in aerospace physiology, a loss of consciousness occurring from excessive and sustained g-forces draining blood away from the brain causing cerebral hypoxia), cortical deactivation through the use of high-power magnetic fields, mystical experiences induced through hyper-ventilation, and brain damage caused by surgery or strokes. If this is true, then neural activity not only fails to be a sufficient condition for mental activity but it may not even be a necessary condition.

Reason #3. Another peculiarity in this discussion is that something like pain, for instance, can be induced in a physical way, but there is no evidence that experimental stimulation of specific neuronal areas is able to produce a specific mental state, let alone a specific thought. Even if every mental event is associated with a brain event, not every brain event makes for a mental event. The presumed jump from matter to "thinking matter" appears to be enormous. The neural system is a necessary, but arguably not a sufficient explanation of thinking. As the late philosopher Mortimer J. Adler emphasized (see chapter 7), there is a clear difference between perceptual and conceptual thought. Thinking in concepts requires universality, whereas sensorial information is only about particulars. The fact that the brain is only a necessary, but not a sufficient, condition of *conceptual* thought, indicates that an *immaterial* intellect is required besides in order to provide an adequate explanation of conceptual thinking. We can even conceptualize what we cannot visualize—something like a circle with four dimensions—which calls for something mental, not neural. Very often we do not see what receptor cells and neuron cells tell us to see, but rather what we wish to see, which is then also not a neural but a mental issue.

Nowadays there is also growing evidence that the neuro-circuitry of the brain is not as static and unchangeable as has long been thought—which is called brain plasticity. This gave Dr. Norman Doidge the idea that the brain can change itself. But that would be sheer magic. Thomas Aquinas would say that whenever something undergoes change, something must be causing that change, for nothing can be the cause of its own change. Therefore, whenever something changes, this change must have been brought about by something other than itself. As nothing can cause itself to exist, so nothing can cause itself to change. The brain could be changed by the mind, but not by itself, for every change requires a cause. However that would only be possible if the mind is not identical to the brain. The brain cannot rewire itself; it needs something else, the mind, which may cause it to rewire.

Reason #4. The German philosopher Gottfried Leibniz once suggested picturing the brain as so much enlarged that one could walk in it as if in a mill.[15] Inside we would only observe movements of various parts but never anything like a thought. For this reason he concluded that thoughts must be different from physical and material movements and parts. Nowadays the mechanical model of cogs and wheels, which Leibniz used, has been replaced by a model of neural and biochemical pathways, but the outcome is still the same. How can an assemblage of neurons—a group of material objects firing away—have any *content*?

Alvin Plantinga describes the problem in all clarity: "What is it for this structured group of neurons, or the event of which they are a part, to be related, for example, to the proposition *Cleveland is a beautiful city* in such a way that the latter is its content? A single neuron (or quark, electron, atom, or whatever) presumably isn't a belief and doesn't have content; but how can belief, content, arise from physical interaction among such material entities as neurons?"[16] Then he says: "Propositions are also mysterious and have wonderful properties: they manage to be about things; they are true or false; they can be believed; they stand in logical relations to each other. How do they manage to do those things? Well, certainly not by way of interaction among material parts."[17]

Whereas the brain as a material entity has characteristics such as length, width, height, and weight, the mind does not have any of those; thoughts are true or false, right or wrong, but never tall or short, heavy or light—they have no mass, no size, no color. If the mental were the same as the neural, thoughts could never be right or wrong and true or false. We can think about sizes and colors of things, but those thoughts themselves do not have sizes and colors. Just as the brain cannot distinguish between legal and illegal narcotics, so is the brain incapable of

15   Gottfried Leibniz, "Monadology 17," in *Leibniz Selections*, ed. Philip Weiner (New York: Charles Scribner's Sons, 1951), 536.
16   Alvin Plantinga, "Against Materialism," 14.
17   Ibid., 21.

telling false thoughts apart from true beliefs. To evaluate the outcome of neural states as true or false, we need something that is not neural. To think differently is like saying that Shakespeare's thoughts were nothing but ink marks on paper.

Reason #5. It is hard to equate the working of the mind with the working of a machine such as a computer—a popular model for the brain. Computers require a human maker and would still need a human subject to give their informational output some meaning or sense. Without human subjects, computers cannot "think." So computers cannot explain the human mind; to the contrary, they must presume its existence. Computers do not create thoughts, but they may carry thoughts that were created by the mind of a human subject—the programmer(s) of the computer program. Consider a voice recognition system: it doesn't really understand what it is programmed to "recognize." Computers only do what we human beings with a mind make them do, for we have proven to be champion machine builders. So the popular slogan "Man versus Machine" is actually very deceiving; it should be "Man versus Man"—Man versus the Man who designed the machine.

Besides, even if a computer may play chess better than Kasparov or any other champion, it plays the game for the same "reason" a calculator adds or a pump pumps—the reason being that it is a machine designed for that purpose—and not because it "wants" to or is "happy" to do so. Computers and similar devices do not have meaning or sense in themselves until a human subject uses them as carriers of information that receives sense and meaning from a human subject. This makes it hard to use the computer analogy to fully understand the human mind, for without the human mind there would be no computers. For those who like to use computer metaphors we might suggest the following: the mind uses the brain as a programmer uses a computer. The same computer can actually run various programs—and so can the brain.

Reason #6. Denial of the human mind is self-destructive. If the mind were just the brain, its thoughts would be as fragile as the molecules they are supposedly based on. It would be sitting on a "swamp of molecules," unable to pull itself up by its bootstraps. Sociobiologists, for instance, claim that we believe what we believe because what we call "truth" emerges from brains shaped by natural selection. But claims like these work like a boomerang—if they are true, they become false. The snake of this claim is swallowing its own tail, or rather its own head, as has been said already in a similar context.

It is actually very hard to deny the mental, because denying the existence of mental activities is in itself a mental activity, and thus would lead to contradiction. Ironically, one cannot deny the mental without affirming it. We mentioned earlier how J.B.S. Haldane and C.S. Lewis have worded this paradox along the following lines: if I believe that my beliefs are the mere product of neurons, then I have no reason to believe my belief true—therefore I have no reason to believe my beliefs are the mere product of neurons. So if we are looking for a key to understanding ourselves it will not be in terms of matter and brain but in terms of mind and soul. The brain is governed by laws of physics, chemistry, and biology; but thoughts are not. As Stephen Barr pithily puts it: "We do not infer the existence *of* our minds, rather we infer the existence of everything else *with* our minds. To put it another way: the brain does not infer the existence of the mind, the mind infers the existence of the brain."[18]

Reason #7. In order to make any mental claims, especially so in science, such claims need to be validated as being true, for otherwise they are worth nothing. If Watson and Crick, or Planck and Einstein, or Darwin and Dawkins, or any other scientists, were nothing but their neurons, their scientific theories would be as fragile as their neurons. That would be detrimental to their claims and to their status as experts in their respective fields. If our mental activities are only the by-product of neural events, they could be nothing more than illusions, or at best mere sensations. If we were nothing but a "pack of neurons," this very statement would not be worth more than its molecular origin, and neither would we ourselves, who are making such a statement.

It requires a mind to come up with generalizations and abstractions such as the law of gravity. Newton's mind, for instance, was able to see beyond the sensorial impression of a falling apple. Hence it has also been argued that the infinite universal meaning of an abstract concept cannot be inscribed in the finite material system of the brain. Besides, the physical world can never be studied by something purely physical, any more than neurons could ever discover neurons. The *knowing* subject must be more than the *known* object, for it requires a mind to understand the

18 Stephen Barr, *Modern Physics and Ancient Faith*, 188.

brain, and it requires a subject to study any object. To explain the mind in terms of physics obscures the fact that one would still need to have a mind first before one could even have physics. While idealists such as George Berkeley reduce reality to "mind" and deny the existence of "matter," materialists reduce reality to "matter" and deny the existence of "mind." But on what grounds do they assume reality must be of only one kind of being? Why cannot reality consist of both matter and mind, of both the mental and the neural?

Reason #8 can be found in information theory and information technology. Crucial in information theory is the separation of content from the "vehicle" that transports it. No possible knowledge of the computer's materials can yield any information whatsoever about the actual content of its computations. The brain is supposed to work in the same way a computer operates since both use a binary code based on ones (1) and zeros (0); neurons either do (1) or do not (0) fire an electric impulse — in the same way as transistors either do (1) or do not (0) conduct an electric current. But that is where the comparison ends. Whatever is going on in the brain — say, some particular thought — may have a material substrate that works like a binary code, but this material substrate only acts as a physical "carrier" for something immaterial: thoughts. Firing neurons are merely carriers of something immaterial such as a thought.

As a consequence, the doctrine of materialism itself has become completely meaning-less. There is no way left to even think in terms of true or false, for as Stephen Barr notes, "One pattern of nerve impulses cannot be truer or less true than another pattern, any more than a toothache can be truer or less true than another toothache."[19] Concepts, thoughts, or beliefs are not material structures. Whatever is going on in the brain — say, some particular thought — may have a material substrate that works like a binary code, but it would not really matter whether this material substrate works with impulses, as in the brain, or with currents, as in a computer, or with letters, as in a book, for the simple reason that this material substrate only acts as a physical "carrier" for something immaterial — thoughts and concepts, that is — coming from the mind, not the brain.

As was said earlier, thoughts are *about* something mental, about something beyond themselves. To use an analogy, anything that shows up on

19   Ibid., 197.

a computer monitor remains just an "empty" collection of "ones and zeros" that do not point beyond themselves until some kind of human interpretation gives sense and meaning to the code and interprets it as being *about* something else. The same holds for firing neurons. Think of what we call a picture: a picture may carry information, but the picture itself is just a collection of tiny dots; it is merely a piece of paper that makes "sense" only when human beings interpret the picture as being *about* something. The same with books: they provide much information for "book worms," but to real worms they have only paper to offer. This means that the neural carrier of information cannot be the same as the mental information it carries.

Reason #9. As stated under reason #3, it does not seem very likely that thoughts can be induced in a physical way. We are not talking here about something like emotions or feelings (even animals have those), because those are physical and biological phenomena that can be physically induced by stimulation of certain brain areas or the use of chemicals. Neither are we referring here to memories stored in the brain—including memories of thoughts once produced by the mind—because memories can be physically stored, similarly to the way thoughts can be "stored" on paper.

Thoughts, on the other hand, cannot be produced in a physical manner, neither by chemicals nor by electrodes. If we could change beliefs with chemicals, presidential candidates would be wise to use that method in their campaigns. If the thought of "two to the power of two" would physically produce the thought of "four," we could have skipped much work in school. Mathematics, for instance, is not something hard-wired in the brain. Unlike brain processes, which are subject to physical causation, thoughts are subject to mental causation based on reason and intellect, on laws of logic and mathematics.

In addition, one could argue that the brain is as much responsible for thinking as is the hand for grasping or the leg for walking. Many of the discussions about the brain's causality of thought seem to involve the idea that if one makes the brain responsible for thought, then it would somehow become the principal agent of thought. But this is as dubious as thinking that if one makes the hand responsible for grasping, somehow it is the principal agent of grasping, as opposed to being a mere tool used by a human being. Instead it might be argued that it is the mind that uses the brain as its organ. The mind needs the brain to function properly, but the brain also needs the mind to function fully. In other words, the neural system may be a necessary, but arguably it is not a sufficient explanation of thinking.

Reason #10 is based on a thought experiment the philosopher Ludwig Wittgenstein once suggested.[20] Picture yourself watching in a mirror how a scientist is studying your opened skull for "brain waves." It can be stated that the scientist is observing just one thing, outer brain activities, whereas the "brain-owner" is actually observing two things — the outer brain activities in the mirror as well as the inner thought processes that no one else has access to. In order to make the connection between "inner" mental states and "outer" neural states, scientists would depend on information that only the "brain-owner" can provide. The world of the mind is only accessible to the "brain-owner." This is so even in court, in spite of lie detector tests. As a matter of fact, very often the only ones to know whether they committed the crime or not are the ones being prosecuted. What lie detector tests may detect are not thoughts, but at best physiological and emotional responses to those thoughts.

It seems obvious that there is no such thing as mind-reading through brain scans or other techniques. Contemporary neuroimaging techniques make it possible only to observe directly the effects of neurological activity such as changes in intracranial blood flow. One cannot "see" cognitive activity itself, but only the effects of cognitive activity. Consequently, neuroscientists cannot just "read" a person's mind. If they want to associate certain brain activities with certain mental activities they need to ask their patients what they were thinking. Hence material explanations cannot possibly lead to a full understanding of non-material phenomena. No mental event is physical, and no physical event is mental. Some speak therefore of a "third-person ontology" versus a "first-person ontology."[21]

In this latter terminology, neural phenomena have a "third-person ontology" whereas mental phenomena have a "first-person ontology," being essentially subjective or "private," directly accessible only to the subject undergoing such mental experiences. No matter how we look at it there seems to exist in this Universe a dualism of properties — neural versus mental, objective versus subjective. The mind has distinct

---

20 Ludwig Wittgenstein, *The Blue and Brown Book* (New York: Harper & Row, 1980), 11–13.

21 E.g., John Searle.

features — such as intimacy, privacy, first-person perspective, and unity of conscious experience — that cannot be found in the brain and its overt, public, third-person ontology. It is ultimately the "I of the beholder" that allows the human mind to study the human brain.

Let us come to a conclusion. If one of the previous ten arguments convinces you completely, that is enough. If they convince you only partially, perhaps all of them together provide enough compelling reason for you to reject the idea that the mental is identical to the neural, that the mind is identical to the brain, or that the soul is identical to the body. Amazingly, many neuroscientists seem to have never heard of these ancient and modern arguments against their strongly held convictions. Or, having heard of them, they have either forgotten or willfully discarded them. They still think of the brain as an organ that secretes thoughts in a way similar to how the hypothalamus secretes hormones.

It always amazes me how some Darwin fans like to downgrade the human mind while touting their own minds. If thoughts were merely the product of bodily and other natural actions, all thoughts would be equivalent, and we would have no way of telling true from false, or knowledge from error. The thought of Darwinians would be worth as much or as little as the thoughts of their opponents. That takes us back to Darwin's question as quoted at the beginning: Can science as practiced by a mind science has found to be itself the result of natural selection be trusted? Again, the answer is no — unless we are ready to derive the human mind (including its rationality and intellect) from a higher source. Some like to call the human mind the human soul; that's fine, as long as we realize that the mind is the intellectual part of the soul. I would say the mind is the soul's eye, its light — or put differently, the mind is the power of the soul by which we know truth. We have definitely entered meta-physical territory here — unfortunately located on a "blind spot" in Darwin's mind.

Darwin could have cleared up the confusion he had created for himself if he could have just acknowledged that the human mind is not a product of natural selection. The human brain (including its intelligence) may well be a product of natural selection, but that doesn't mean the human mind (including its intellect) is too. As a matter of fact, the theory of natural selection must assume the human mind; but it can neither create nor explain it. We need a mind to study the brain, for no brain can study itself. And so the mind must have another origin than the brain. I would even go further and claim that the mind must be something made in God's image, a derivative of the Creator's mind. Whereas it was Darwin's conclusion that due to natural selection we cannot trust anything we know about God, I would rather argue a larger claim — that we cannot trust anything we know at all if there is no God.

Neo-Darwinians are not only gradualists but are likewise brain-mind-equalizers. In their view all so-called uniquely human faculties of language, rationality, morality, self-awareness, and religion can only be traced back to the brain. Therefore Neo-Darwinians keep searching for mutations that have shaped the brain so that in time it became "human." But if the human mind is more than the brain, then uniquely human faculties may not be tied to the brain but to the mind. They would no longer be determined by anatomy, physiology, and genetics, but instead come from the human soul — with the human mind being the intellectual part of the human soul.

If there is indeed a deep divide between the mental and the neural, this divide would place the discussion of what happened to the first humans on a completely different level. It would greatly limit the impact of mutations at the dawn of humanity. The origin of the human mind could then very well come from a completely different source. Let's see how we can get there.

IMMORTAL SOULS

According to Thomas Aquinas, all living beings have a soul (*anima* in Latin). "Soul" is how he explains life. "Soul" is definitely a concept in the sense of concepts we described before: it has a meaning but also a reference which is invisible (see chapter 7). When someone wants to know why an apple falls to the ground we mention gravity, which is another concept with an invisible reference; gravity explains why things fall. In the same way, when someone asks why an organism is alive while a rock is not, we could mention the soul. "Soul" explains why things are alive. However, when a plant or animal dies, its soul goes with it.

More specifically, Aquinas speaks of a "vegetative" or "nutritive" soul for plants, a "sensitive" soul for non-human animals, and an "intellective" or "rational" soul for humans. This may sound like antiquated philosophy, since overtaken by modern science, but that may be deceiving. It is actually common-sense philosophy. Plants have a vegetative soul, with growth, metabolism, and reproduction; animals have a sensitive soul, with locomotion and perception; and human beings have a rational soul, with reason and intellect. These are "nested" in the sense that anything possessed of a higher degree of soul has also all the lower degrees. All living things grow, nourish themselves, and reproduce. Animals not only have those activities, but also move and perceive. Humans also use reason, but do all of the above as well. Aquinas's ancient views still nicely tie in with current scientific insights.

All of this is based on the Thomistic (and Aristotelian) idea that every being is a unity of *matter* and *form*. Each and every individual thing is composed not only of matter [*materia*] but also of form [*forma*]. These two cannot exist without each other. There are no physical beings without matter, and there are no beings without form. Chesterton explained this as follows in almost paradoxical terms: "Matter is the more mysterious and indefinite and featureless element; and ... what stamps anything with its own identity is its Form. Matter, so to speak, is not so much the solid as the liquid or gaseous thing in the cosmos; and in this most modern scientists are beginning to agree with him. But the form is the fact; it is that which makes a brick a brick, and a bust a bust, and not the shapeless and trampled clay of which either may be made."[22]

But if all life is outfitted with soul, what makes a human soul so different? What causes the soul divide of humanity, if there is one? As a matter of fact there is an important difference between the three types of souls. Vegetative and sensory souls depend entirely on matter for their operation — for instance, photosynthesis requires leaves and digestion is done in intestines. However, whereas plant souls and animal souls are coextensive with matter, the human soul has to exceed matter because it is able to grasp and process abstract ideas such as "circle," "truth," or "justice," which do not have length, width, or height. The human intellect is essentially *immaterial*, not requiring any bodily organ for its operation. We do need sensory organs for perception and sensation — to see a particular circle, for instance — but thinking about circles goes far beyond that one particular circle we are looking at. Perceived objects are never

---

22 G.K. Chesterton, *St. Thomas Aquinas* (New York: Sheed and Ward, 1933), chapter VI ("The Approach to Thomism").

perfect circles, but the thought or concept of a circle is about a perfect circle, without its being an actual circle itself. Any material circle is always and only an approximation of perfect circularity. Therefore the thought of circularity cannot be itself material. This makes the human intellect, whose activity is not the activity of any bodily organ, essentially *immaterial*. Thoughts are universal and immaterial, whereas perceived objects are particular and material. Neurons may "carry" thoughts, but they do not produce thoughts.

To put it another way, the human soul has the capacity to grasp the complex and abstract ideas that we can express in human language (see chapter 6). The Catholic Church confirms this: "The unity of soul and body is so profound that one has to consider the soul to be the 'form' of the body: i.e., it is because of its spiritual soul that the body made of matter becomes a living, human body." [23] Since a soul is "fitted" to its body in the same way that a key is fitted to its lock, a human soul can only "in-form" a body that has a brain complex enough to deal with and process language, rationality, morality, self-awareness, and religion. These faculties do not seem to come from the body but rather from the soul. They are essentially immaterial, not requiring any bodily organ for their operation. They are not matter but form instead. If the intellect were really a material thing — say, part of the brain — then that thing would become a circle whenever we think of circles and circularity, which is absurd. The operations of the intellect cannot consist of purely material processes. Instead, it is the form, not the body, that makes us human. It is the form that comes with the faculties of language, rationality, morality, self-awareness, and religion. Since immaterial things have, unlike material things, no natural tendency to decay, the soul does not go out of existence when the material body does.

This might be seen as an inconsistency in Aquinas' philosophy. On the one hand he says that when a plant or animal dies its soul goes with it. On the other hand he seems to make an exception for human beings. Because Aquinas declares that body and soul make a tight unity and therefore cannot exist without each other, it seems to follow that the human soul cannot possibly exist without the body either. Aquinas readily admits this is true when he says, "It is clear that the soul is united

23 *Catechism of the Catholic Church*, 365.

to the body by nature: because by its essence it is the form of the body. Therefore it is contrary to the nature of the soul to be deprived of the body."[24] But does this necessarily mean that the soul cannot exist on its own after death? Let's see why this does not follow.

Aquinas regarded the human soul as independent of the body in its existence, so it can "subsist" even after the dissolution of the body of which it is the "form." This means that the human soul has its being and its operation in itself, independent of anything else, including the body—for its intellectual acts are not the acts of any material organ. According to Aquinas a human person is a material substance with an immaterial part, the soul; but this immaterial part is a substance in itself, because the soul's intellectual operations can carry on independently of the body. This makes Aquinas conclude that whatever can operate independently must *exist* independently. Therefore the human soul is capable of existing apart from the body, which makes it possible for the human soul to survive death, that is, the cessation of life in the body.

This idea of an independent soul, in spite of its connection with matter, the body, may teeter on the verge of contradiction, but it is not as strange as it sounds. Think of the analogy of language: We think with and through language, yet our thinking surpasses language and is not absorbed by it in spite of our dependence on language. Dependence does not imply identity; dependence is one thing, identity quite another. In a similar way, although an embodied soul is fully tied to matter, it also surpasses matter and is not absorbed by it. Therefore its capacity for immortality becomes a reality. It seems to make perfect sense that if the soul persists while the body keeps changing, the soul may also persist after the body has changed so much that it is corrupted by death.

Yet, although reason tells us humans are mortal—even if their souls are immortal—it is contrary to the nature of the soul to be deprived of the body. It is not a complete substance, as it is for Descartes, but an *incomplete* substance (think of a heart that temporarily subsists on its own after having been taken from a donor). Consequently, says Aquinas, when separated from the body, the soul is in a "violent" state, and the desire of the immortal soul is to be joined again with its own body. This is not required by human nature as such, since by nature all humans are

24 Aquinas, *Summa contra Gentiles*, IV, 79.

mortal—that is, upon death "they die," thus ending their being a person, even though their souls do not die—and therefore the resurrection of the body always remains a gratuitous gift of God to make the person whole again. Although the immaterial soul survives a person's bodily death, the *person* that one is must necessarily be embodied.[25] Terminology is important here: the unity of a human body and a human soul is a *person*.

Obviously, for human reason, death remains an impassible barrier: Human beings cannot conquer death even if they are able to escape it through their souls. After death, a separated soul does not depend on the body for its existence, but there is no longer a (full) *person*. Because the human, spiritual, and immortal soul is naturally incomplete as subsisting apart from the body, Aquinas sees the state of a separated soul as unnatural for it, and an opening for—but not a compelling argument, of course, for—the resurrection of the body, when body and soul will be united again into one *person*. In the words of the Catholic Church: "The human body shares in the dignity of 'the image of God': it is a human body precisely because it is animated by a spiritual soul, and it is the whole human *person* [italics added] that is intended to become, in the body of Christ, a temple of the Spirit."[26]

Some may object that this soul-talk comes very close again to a "ghost in the machine" kind of philosophy (see chapter 9). When Ryle coined this phrase he was not referring to Aquinas but was actually rejecting the dualism of René Descartes—the dualism of body and mind. When Descartes compared our minds with a pilot in his ship, he seemed to suggest that the mind is the pilot behind everything the body does. One of the problems of Cartesian dualism is that it cannot explain how body and mind could possibly *interact*. Not only are minds and bodies different entities, but they are radically different sorts of entities. How could a mind, a non-spatial item, ever cause effects in a spatial item, the body? Besides, in this Cartesian view, a pilot can be without a ship, and a ship can be without a pilot, but in the Thomistic view, there is no human body without a human soul (unless it is a corpse), and there is no soul without a body (except temporarily, until the resurrection after death).

25 Aquinas, *Summa contra Gentiles*, IV, 79.
26 *Catechism of the Catholic Church*, 364.

However, Aquinas would not agree with Cartesian dualism. He would place the issue of body and mind in the wider context of the relationship between body and soul—with the mind being the intellectual part of the soul. In Aquinas's view, body and soul are not two complete substances, as in Cartesian dualism, but two components of one complete substance, the human person. They make for a unity whose nature is comprised of both material *substance* and immaterial *form*, so that the body becomes what it is due to the soul (and its mind), for matter can never exist all by itself. The soul is not taken by him to be "made up" out of anything—it is not some kind of "stuff," it is not even made out of "spiritual matter" (whatever that would be).

As a consequence, a human *person* must be seen as a unity of body and soul. In a sense, the body depends on the soul and the soul depends on the body. But their unity, their combination, cannot be explained by either one, for that would lead to a vicious circle. A person, as a unity of body and soul, cannot be the cause of body and soul being unified, for that would mean a person is both the cause and the effect of a body and soul—which is incoherent. Nothing can be the cause of itself, lifting itself up by its own metaphysical bootstraps. Therefore, the unity of body and soul in a person must come from something outside them that accounts for their combination—which could be seen as an argument for the existence of God.

Biology, on the other hand, tends to isolate the body from the person, in spite of the fact that the human body is always a person's body with a "self" (see chapter 9). And the same should be said about the soul—it is always a person's soul. Unfortunately, Descartes's view has strongly and profoundly permeated our culture by driving a wedge between body and soul, usually at the cost of the soul. We tend to separate them and set them in an antagonistic relationship to each other—a master/slave relationship. The body is often seen as a prison for the soul from which the soul wants to escape. But Aquinas sees things very differently: When separated from the body, the soul is in a "violent" state, and the desire of the immortal soul is to be joined again with its own body. When you die, your soul carries on, but you yourself do not. However, the survival of your soul makes it possible for you as a person to live again, God willing.

It is obvious from this that a human soul is a philosophical and theological concept with a very specific meaning and reference. The fact that it is not a scientific, material concept does not mean it is not a legitimate concept, as we have seen earlier. Where, then, does the human soul come from? Similarly to what we said already about the origin of rationality and morality, the human soul does not come from evolution or genes, but from God. Matter can evolve, but since the human soul is immaterial and cannot evolve, it has to be created directly by God. And so, since science deals only with matter—with what can be measured, counted, quantified, and dissected—the origin of the human soul eludes science. The Catholic Church summarizes this as follows: "The Church teaches that every spiritual soul is created immediately by God—it is not 'produced' by the parents—and also that it is immortal: it does not perish when it separates from the body at death, and it will be reunited with the body at the final Resurrection."[27]

Is all of this mere speculation on Aquinas's part, or is there some neuroscientific evidence that the soul does have the capacity to persist after the body breaks down? A possible confirmation might come from so-called near-death experiences (NDEs)—a broad range of personal experiences associated with impending death, encompassing sensations such as detachment from the body, feelings of levitation, total serenity, security, warmth, the experience of absolute dissolution, and the presence of a bright light. One of the first clinical studies of NDEs in cardiac arrest patients was done in 2001 by Pim van Lommel, a cardiologist in the Netherlands.[28] With his team, he studied a group of Dutch patients who had been brain-dead from cardiac arrest but were successfully revived. Of the 344 patients who were successfully resuscitated after suffering cardiac arrest, sixty-two experienced "classic" NDEs, which included out-of-body-experiences. Of these sixty-two patients, 50 percent reported an awareness or sense of being dead, 24 percent said that they had had an out-of-body experience, 31 percent recalled moving through a tunnel, and 32 percent described meeting with deceased people. Van Lommel concluded that his findings supported the theory that consciousness had continued despite lack of neuronal activity in the brain—a flat EEG. What such experiences suggest is that the mind and soul can survive brain death.

27 *Catechism of the Catholic Church*, 366.

28 P. Van Lommel, et al., "Near-death experience in survivors of cardiac arrest: a prospective study in the Netherlands," *Lancet*, 358 (2001): 2039–45.

No wonder NDEs are often cited as evidence for the existence of the human soul. On the other hand, not surprisingly, all kinds of *biological* explanations have been suggested in reply: oxygen deprivation (anoxia), high carbon monoxide levels, REM-sleep phenomena, psychedelic agents, hallucination. Obviously, the question remains why not all people under those circumstances had NDEs. However, further research has been done to rule out these explanations. A recent study by Dr. Sam Parnia suggests that NDE patients are "effectively dead," having no neural activities that would be necessary for dreaming or hallucination. Additionally, in order to rule out the possibility that NDEs resulted from lack of oxygen, Parnia rigorously monitored the concentrations thereof in the patients' blood and found that none of those who underwent the experiences had low levels of oxygen.[29] He was also able to rule out claims that unusual combinations of drugs were to blame because the resuscitation procedure (and thus the drugs involved) was the same in every case, and not every patient had an NDE.

As stated many times earlier, the soul is not a (neuro-) scientific concept. It is Aquinas's philosophy that shows us that the human soul, which gives form to the body, is an entity that subsists by itself (*per se*), and not in virtue of some other reality such as the body. Therefore it has not only its own reality but also its own independence from the body, to which it is substantially united in such a way that it forms, along with the body, one being—a person. And so we see that a human being is not a purely material being but a composite of spirit and matter—an embodied spiritual being. However, the spirit (or soul) subsists *per se* and therefore is able to exist without the body (or flesh) after death. In other words the soul is that part of us that lasts forever. At the resurrection we do not become someone other than the very same self that was once begotten and born into this world.

ADAM AND EVE?

The previous considerations have led us to an astounding conclusion: That which makes a person a human being is a human soul. If that is true, then we have to face a daunting question: how and when did the human

---

29   Sam Parnia, et al., "AWARE—AWAreness during REsuscitation—a prospective study," *Resuscitation*, 85 (2014): 1799–1805.

soul emerge individually in the course of evolution? If the human soul cannot be reduced to something merely material, then obviously it cannot be generated by merely physical processes such as sexual reproduction, evolution, or genetics. Since a soul does not admit of degrees — you either have one or you don't — in our ancestry there must have been a first creature, or set of creatures, endowed with an immortal human soul and the mental powers of language, rationality, morality, self-awareness, and religion, which come with a human soul. You either have those faculties or you don't. Nor do they come in degrees.

Obviously, the bodies of human creatures must have been suited to receive a human soul, including for this purpose the right genes and the proper brains. We referred to such conditions already when we discussed when and where the human faculties of language, rationality, morality, self-awareness, and religion emerged — probably somewhere in Africa some 80,000 years ago.

What happened then and there in Africa is hard, if not impossible, to describe in scientific terms, because we are dealing with unique human faculties of an immaterial nature. On the one hand Thomas Aquinas argues that a certain kind of body is necessary, but not sufficient, for these unique faculties. On the other hand these unique faculties require something beyond the power of any bodily organ — something, therefore, that can only come into being, in each individual case, through a creative act of God. And next, what comes with the human soul are the faculties of language, rationality, morality, self-awareness, and religion. *The Catechism of the Catholic Church* emphatically confirms this: "The soul, the 'seed of eternity we bear in ourselves, irreducible to the merely material,' can have its origin only in God."[30] At one point Pope John Paul II declared emphatically: "Evolution does not suffice to explain the origin of the human race, just as the biological causality of the parents alone cannot explain a baby's birth."[31]

We discussed earlier (see chapter 5) that there is the issue whether humanity originated from one or two individuals (monogenism) or from a pool of individuals (polygenism). The Catholic Church is very definite

---

30 *Catechism of the Catholic Church*, 33. The citation used inside the quote is from *Gaudium et Spes*, 18 §1.

31 John Paul II, general audience (May 27, 1998), "Spirit Enables Us to Share in Divine Nature," *L'Osservatore Romano* (English), June 3, 1998, n. 5.

in defending monogenism—not for biological reasons but for theological reasons—more specifically, reasons based on the doctrine of Original Sin. This doctrine is pivotal in Catholic Faith, for if there is no Original Sin, the Cross is a hoax; and if there is no Cross, the whole economy of Salvation through the Incarnation is up for grabs. The *Catechism* puts it this way: "The Church, which has the mind of Christ, knows very well that we cannot tamper with the revelation of original sin without undermining the mystery of Christ."[32] Clearly, if humanity is not in fleshy solidarity with itself due to a descent from a common ancestor, then Christ's fleshy solidarity with all of humanity is rendered problematic.

In his 1950 encyclical *Humani Generis*, Pope Pius XII minced no words: "No Catholic can hold that after Adam there existed on earth true men who did not take their origin from him as the first parent of all, or that Adam is merely a symbol for a number of first parents."[33] Blaise Pascal said it well: "It is dangerous to show man in how many respects he resembles the lower animals, without pointing out his grandeur." But Pascal then emphatically added: "It is also dangerous to direct his attention to his grandeur, without keeping him aware of his degradation"[34]—which is a reference to the Fall. In the words of Pope Pius XII: "Only from a man could there proceed another man who would call him father and progenitor."[35]

As a matter of fact, the human race forms a unity—not only in its origins, but also in its sinful state as well as in its salvation. The descendants of Adam were distorted by sin because Adam, who was their source, chose to distort himself. How do we understand this in the light of evolution? The answer would be: the first humans, Adam and Eve, were given the faculties of rationality and morality, but they deliberately chose to follow their old animal drives.

What is the truth of all of this? Truth is truth, whether you like it or not, whether you believe it or not, whether you know it or not. The ultimate truth is that truth cannot be established by a majority vote. However, there are two kinds of truth. There are truths that we discover and

32 *Catechism of the Catholic Church*, 389.
33 Encyclical *Humani Generis*, 37.
34 Pascal, *Pensées*, 418.
35 Pius XII, in an allocution to the Pontifical Academy of Sciences, *Acta Apostolicae Sedis*, XXXIII (1941), 506.

know with the help of *reason*, but there are also truths that we can only discover and know with the help of *faith*. Another way of distinguishing these two kinds of truth is this: There are "natural" truths that come to us by experience or experiment and "supernatural" truths that come to us through Revelation and Scripture; or also, "human" truths that we gain on our own and "divine" truths that we receive from Heaven; or also, truths we can see through our physical eyes and truths we discern through the eyes of faith.

The same distinction can be made when it comes to science and religion: Science has *theories* to help us understand the world better and better, but they are subject to change — so let us not make science more than what it is. Religion on the other hand has *dogmas* that we try to understand more and more, but they never change — so let us not make religion less than what it is. Truth is truth, but we may not yet fully understand the truth. What is true of Catholic doctrine today will also be true tomorrow. But not so in science: Science is always an ongoing process, perhaps even never-ending; what is true today may have to be revised tomorrow based on new or better tests and experiments. Scientists must submit their minds to the data of experiment, but religious believers must submit theirs to the data of God's Revelation.

In spite of this distinction, Thomas Aquinas saw with utter clarity that *all* truth comes from God, and therefore that ultimately there can never be any conflict between the findings of *reason* and the beliefs of *faith*, or between the data of the sciences and the facts of revelation, or between scientific truths and theological truths. He claimed that faith and reason, or theology and philosophy, play, in his own words, "complementary roles in the quest for truth. Grace does not destroy nature but fulfills it."[36] Because God is the Author of both kinds of truth, they cannot contradict each other. Contradictory claims can't be simultaneously true. Therefore, if we find a seeming contradiction between the two, we have not correctly understood either the Scriptures or the natural world, or both.

In our search for the origin of humanity, we have a case like this. Polygenism may be the latest, provisional theory in science, but monogenism is what divine revelation and religious faith tell us to be the truth. When we detect a contradiction, either one or both truths must have

36 Aquinas, *Summa Theologiae*, part 1, question 1, article 8, response to objection 2.

been claimed in error and must be seriously evaluated. The *Catechism* summarizes all of this very nicely: "Though faith is above reason, there can never be any real discrepancy between faith and reason. Since the same God who reveals mysteries and infuses faith has bestowed the light of reason on the human mind, God cannot deny himself, nor can truth ever contradict truth."[37] Since Revelation tells us that human nature has been deeply corrupted from its very beginning (the Fall) and that Jesus's crucifixion was needed to redeem us from this universal corruption, monogenism is a theological truth that forces us to reexamine any current scientific "truths" that contradict it.

Now back to the question: how did humanity originate in the process of evolution? As we showed in the previous chapters, all the "preparatory work" for the dawn of humanity had to be done with smaller or larger steps, all of which is part of a biological process. Apart from that, there was also a "spiritual process" in which God chose to raise two members of a population of "anatomically modern humans" to the spiritual level of language, rationality, morality, self-awareness, and religion. Pope John Paul II spoke of "an ontological leap" at the moment of transition to the spiritual:

With man, we find ourselves facing a different ontological order—an ontological leap, we could say. But in posing such a great ontological discontinuity, are we not breaking up the physical continuity which seems to be the main line of research about evolution in the fields of physics and chemistry? An appreciation for the different methods used in different fields of scholarship allows us to bring together two points of view which at first might seem irreconcilable. The sciences of observation describe and measure, with ever greater precision, the many manifestations of life, and write them down along the time-line. The moment of passage into the spiritual realm is not something that can be observed in this way—although we can nevertheless discern, through experimental research, a series of very valuable signs of what is specifically human life. But the experience of metaphysical

37 *Catechism of the Catholic Church*, 159.

knowledge, of self-consciousness and self-awareness, of moral conscience, of liberty, or of aesthetic and religious experience — these must be analyzed through philosophical reflection, while theology seeks to clarify the ultimate meaning of the Creator's designs.[38]

So the search for the "real" Adam and Eve is on. The fact that science has found a *genetic* "Adam" and "Eve" — representing our most recent common matrilineal and patrilineal ancestors (see chapter 5) — does not mean we have to stop searching for the *real* Adam and Eve, who stand at the origin of language, rationality, morality, self-awareness, and religion. But such a search is beyond the scope of science. For this reason Chesterton once acknowledged the possibility that man's "body may have been evolved from the brutes," but then he adds emphatically that "we know nothing of any such transition that throws the smallest light upon his soul as it has shown itself in history."[39] *The Catechism of the Catholic Church* words this more forcefully: "The doctrine of the faith affirms that the spiritual and immortal soul is created immediately by God."[40]

So when we discuss the emergence of humanity in evolution, we need to distinguish the notion of a creature "human" in a strict metaphysical sense from that of a creature which is "human" merely in a biological sense — with a certain brain volume, an erect posture, an opposable thumb, and so on (see chapter 5). The biological creature would be more or less like us in its bodily properties but would lack our spiritual powers. In short, it would lack a human soul and mind. Chesterton again: "There may be a broken trail of stones and bones faintly suggesting the development of the human body. There is nothing even faintly suggesting such a development of the human mind.... We can accept man as a fact, if we are content with an unexplained fact. We can accept him as an animal, if we can live with a fabulous animal."[41] In other words, there is a deep divide between the two. In the words of Pope John Paul II:

---

38 John Paul II, "Message to the Pontifical Academy of Science: On Evolution," Oct. 22, 1996, sec. 6.

39 G. K. Chesterton, *The Everlasting Man*, book 2, part I, chapter 2 ("Professors and Prehistoric Men").

40 *Catechism of the Catholic Church*, 382.

41 G. K. Chesterton, *The Everlasting Man*, part I, chapter 1 ("The Man in the Cave").

"The moment of passage into the spiritual realm is not something that can be observed with research in the fields of physics and chemistry."[42]

As we said earlier, the mind itself is not a biological phenomenon—or to put it briefly, the mind is not the brain—in spite of the fact that many, especially in the scientific profession, remain dedicated "brain-mind equalizers." They just keep ignoring, or even denying, that the mind is different from the brain. Instead, we should say, as the neurologist Viktor Frankl put it, that while the brain conditions the mind, it does not give rise to it.[43] In plain words, the brain does not secrete thoughts like the pancreas secretes hormones. If thoughts were merely a product of the brain, then thoughts about the brain would also be merely a product of the brain. This leads us into a vicious circle. Therefore, the mind must be more than, or at least different from, the brain—there must be more to the mind, which is the intellectual part of the soul, than just the brain, which is a biological part of the body.

Consequently, we didn't inherit our souls from our parents, and our souls certainly didn't come with our DNA, nor did they come through the umbilical cord. Nevertheless, there need not be any conflict between what science tells us and what religion tells us about the emergence of humanity. Science knows nothing about soul and mind; religion on its own knows basically nothing about body and evolution. But together they may tell us the whole story. Sir John Eccles—an Australian neurophysiologist who won the 1963 Nobel Prize in Physiology or Medicine for his work on the neural synapse—relates how his teacher, the Nobel Laureate and neurophysiologist Charles Sherrington, told him five days before he died, "For me now, the only reality is the human soul."[44]

If we didn't inherit our souls from our parents, then our souls didn't come from Adam and Eve either. This means also that biology can't tell us much about our first ancestors. It is important to keep in mind that "mitochondrial Eve" and "Y-chromosomal Adam" are merely the most recent common matrilineal and patrilineal ancestors. We should certainly

---

42 Message delivered to the Pontifical Academy of Sciences, October 22, 1996.

43 Viktor Frankl in a discussion of J.R. Smythies's paper, "Some Aspects of Consciousness," in *Beyond Reductionism*, eds. Arthur Koestler and J.R. Smythies (London: Hutchinson, 1969), 254.

44 John Eccles, *Facing Reality: Philosophical Adventures of a Brain Scientist* (New York: Springer-Verlag, 1970), 174.

not confuse them with Adam and Eve as presented in the Bible. Genetics studies the *Book of Nature* in which "Y-chromosomal Adam" and "mitochondrial Eve" represent branching points in the tree of evolution, whereas the Judeo-Christian religion studies the *Book of Scripture*, in which Adam and Eve represent the first parents of humanity, who started the first dysfunctional family, which developed a broken relationship with God. We cannot identify or equate them; we cannot read the Book of Scripture as if it were the Book of Nature, nor vice versa. The Book of Nature tells us where we come from in a biological sense, whereas the Book of Scripture tells us where we come from in a religious sense.

Pope Benedict XVI summarized all of this beautifully during a conference at Castel Gandolfo in 2008:

> The clay became man at the moment in which a being for the first time was capable of forming, however dimly, the thought of "God." The first Thou that — however stammeringly — was said by human lips to God marks the moment in which the spirit arose in the world. Here the Rubicon of anthropogenesis was crossed. For it is not the use of weapons or fire, not new methods of cruelty or of useful activity, that constitute man, but rather his ability to be immediately in relation to God. This holds fast to the doctrine of the special creation of man...[Herein] lies the reason why the moment of anthropogenesis cannot possibly be determined by paleontology: anthropogenesis is the rise of the spirit, which cannot be excavated with a shovel.[45]

45 *Creation and Evolution: A Conference With Pope Benedict XVI in Castel Gandolfo*, Stephan Horn, S.D.S. (ed.), 2008, 15–16.

A Final Word

THIS BOOK TRIED TO ADDRESS SOME FUNDAMENTAL questions about the dawn of humanity. What can we expect science to tell us about the origin of humanity in evolution? How could it happen that the first humans were able to use language, to think rationally, to act morally, to know who they were, and to know there is a God?

At the dawn of humanity there appeared quite suddenly and abruptly the faculties of language, rationality, morality, self-awareness, and religion, all combined in an immortal soul—practically all at once. In the words of Pope John Paul II: "But if we go down into the depths of man, we see that he is more different from [the rest of] nature than he resembles it. Man possesses a spirit, intelligence, freedom, conscience; therefore he resembles God more than the created world."[1] So we can rightly say with Berwick and Chomsky, "*Only us!*" We are still puzzled, though, as to how to explain all of this. Nevertheless, many people expect science will give us the answers, and will give us even the final answer. Is it realistic to expect science to give us a final answer? There are at least three reasons why science cannot honor such a high expectation.

The first reason is that science never has *final* answers. Science is a work in progress. What we call "proven" scientific knowledge is only proven until a new set of empirical data "disproves" what was previously considered "proven." In science, whatever is true today may be false tomorrow. Francis Crick, one of the two scientists who discovered DNA, couldn't have said it more forcefully: "A theory that fits all the facts is bound to be wrong, as some of the facts will be wrong."[2] He could have expressed this more accurately, though, by stating that facts cannot be wrong, but they may turn out not to be facts.

Nevertheless, some scientists cannot resist the temptation to claim the possibility of certainty and finality in science. But is their dream realistic? Most people would admit that certainty and finality may "not yet"

1   John Paul II, Address to Youth in the Vatican, December 6, 1978.
2   Francis Crick, *What Mad Pursuit: A Personal View of Scientific Discovery* (New York: Basics Books, 1990), 60.

be possible in "softer" sciences such as biology and paleoanthropology, but as a matter of fact, it is not even possible for the "hard" science of physics. The Dutch physicist Pieter Zeeman, later to become a Nobel laureate, was fond of telling how in 1883, when he had to choose what to study, people had strongly dissuaded him from studying physics. "That subject's finished," he was told, "there's no more to discover."[3] It is even more ironic that this also happened to Max Planck, since it was he who, in 1900, laid the foundations for one of the greatest leaps in physics, the quantum revolution.[4] Fortunately, some years ago, Stephen Hawking ended his inaugural lecture with a more realistic note: "though we know what end we are looking for, that end does not seem to be in sight for quite a long while to come."[5] Yet it remains a timeless temptation to claim that the unknown has been reduced to almost nothing. Really? The magnitude of the unknown is, well . . . unknown! Science is by its nature a work in progress. It does not seem to have *final* answers.

The second reason why science cannot honor the high expectations of some of its fans, especially Neo-Darwinians, is the fact that the emergence of humanity in evolution is not an exclusively biological, scientific issue. In several of the previous chapters we left open the possibility that science may develop new insights into the genetics behind the faculties of language, rationality, morality, self-awareness, and religion. However, if these faculties also have a strong connection to having an immaterial and immortal human soul — and we found strong indications that they do — it may be doubtful if genetics, or science more generally, can fully explain how these faculties came along. When we "physicalize" or "biologize" some particular phenomenon, we sacrifice precisely those features that give it a character distinct from physics or biology. Therefore, whatever is missing in a certain scientific description is in fact missing, not because it is absent from reality, but because it is absent from this particular scientific description of reality.

The third reason why science cannot honor the high expectations some

3 Mentioned by the physicist A. Van Den Beukel, *The Physicists and God* (North Andover, MA: Genesis Publishing, 1996), 37.

4 From a 1924 lecture by Max Planck: *Scientific American*, Feb. 1996, 10. See also Alan P. Lightman, *The Discoveries: Great Breakthroughs in Twentieth-Century Science, Including the Original Papers* (Toronto: Alfred A. Knopf, 2005), 8.

5 Stephen Hawking, *Is the End in Sight for Theoretical Physics: An Inaugural Lecture* (Cambridge: Cambridge University Press, 1980), 22.

of its fans entertain is that science must presuppose certain assumptions that it cannot justify on its own. Scientists assume that there is an objective world outside their own minds. They also must assume that this world has a built-in order governed by regularities of the sort captured in scientific laws of nature. They must assume too that the human intellect and its perceptual apparatus can uncover and accurately describe these regularities. And there are other assumptions. But in no way can scientists justify these assumptions without arguing in a circle. Nevertheless, science could never get off the ground without these assumptions.

The fourth reason why science cannot honor the high expectations some of its fans entertain for it is that science cannot have *all* the answers to *all* our questions. Nonetheless, many still believe that all our questions do eventually have a scientific answer phrased in terms of particles, quantities, and equations — in their minds, it's just a matter of time. As Edward Feser eloquently put it: "Since the equations of physics are, by themselves, *mere* equations, *mere* abstractions, we know there must be something that makes it the case that the world actually operates in accordance with the equations."[6] People who do not realize this glorify whatever science studies, whatever can be dissected, counted, quantified, and measured — matter, that is. Their claim is that there is no other point of view than the "scientific," "materialistic" world-view. They believe there is no corner of the Universe, no dimension of reality, no feature of human existence beyond its reach. However, those who believe that only science is able to tackle and answer all our questions have a dogmatic, unshakeable belief in the omni-competence of science and its unrivalled power. This belief is commonly called *scientism*. What is wrong with such a belief?

A first reason for questioning the viewpoint of scientism is a very simple objection: Those who defend scientism seem to be unaware of the fact that scientism itself does not follow its own rule. How could science ever prove all by itself that science is the only way of finding truth? There is no experiment that could do the trick. Science cannot pull itself up by its own bootstraps — any more than an electric generator is able to run on its own power. One cannot talk *about* science without stepping *outside* science. Well, scientism must step outside science to

6   Edward Feser, *Five Proofs of the Existence of God*, 45.

claim that there is nothing outside science and that there is no other point of view—which does not seem to be a very scientific move. Whenever you ignore or neglect something, that does not entitle you to reject it as non-existent. The reason why non-material features, for instance, fail to show up in science is not that the scientific method has shown us they don't exist, but rather that this method has stipulated ahead of time that they be left out of the description.

Consequently, the truth of the statement "no statements are true unless they can be proven scientifically" cannot itself be proven scientifically. It is not a scientific discovery but at best a philosophical or metaphysical viewpoint—and a poor one at that. It declares everything outside science to be a despicable form of metaphysics, in defiance of the fact that all those who reject metaphysics are in fact committing their own version of metaphysics. Scientism rejects any religious faith and replaces it with its own "faith." It uses its own dogma of scientism to reject all religious dogmas. This makes scientism a totalitarian ideology, for it allows no room for anything but itself.

A second reason for rejecting scientism is that a method as successful as the one that science provides does not disqualify any other methods. A blood test, for instance, is an excellent method to assess a person's health, but there are many other reliable methods, such as X-rays, MRIs, etc., depending on what we are trying to assess. But a blood test on its own cannot be used to prove that a blood test is the best and only method there is. Yet that is what scientism does. First it declares one particular method, the scientific method (whatever that is), as far superior and then claims that this superiority disqualifies any other methods. It makes for megalomania, transforming science into a know-all and cure-all. As Edward Feser puts it: "That a method is especially useful for certain purposes simply does not entail that there are no other purposes worth pursuing nor other methods more suitable to those other purposes."[7]

The late University of California at Berkeley philosopher of science Paul Feyerabend, for instance, comes to the same conclusion as Feser when he says that "science should be taught as one view among many and not as the one and only road to truth and reality."[8] Even the "posi-

---

7   Ibid., 282.
8   Paul Feyerabend, *Against Method: Outline of an Anarchistic Theory of Knowledge* (New York: Verso Books, 1975), viii.

tivistic" philosopher Gilbert Ryle expressed a similar view: "[T]he nuclear physicist, the theologian, the historian, the lyric poet, and the man in the street produce very different, yet compatible and even complementary pictures of one and the same 'world.'"[9] As a matter of fact the astonishing successes of science have not been gained by answering every kind of question, but precisely by refusing to do so. One could even agree with the late Nobel Laureate and biologist Konrad Lorenz that a scientist "knows more and more about less and less and finally knows everything about nothing."[10]

A third argument against scientism is that scientific knowledge does not even qualify as a superior form of knowledge; it may be more easily testable than other kinds, but it is also very restricted and therefore requires additional forms of knowledge and additional methods and tools of detection. Consider the analogy used by the philosopher Edward Feser: A metal detector is a perfect tool to locate metals, but that does not mean there is nothing more to this world than what metal detectors can detect.[11] An instrument can only detect what it is designed to detect. Instead of letting reality determine which techniques are appropriate for which parts of reality, scientism lets its one and only favorite technique dictate what is considered "real" in life — in denial of the fact that science has purchased success at the cost of limiting its ambition.

To best characterize this restricted attitude of scientism, an image used by the late psychologist Abraham Maslow might be helpful: If you only have a hammer, every problem begins to look like a nail.[12] So, instead of idolizing our "scientific hammer" we should acknowledge that not everything is a "nail." Even if we were to agree that the scientific method gives us better testable results than other sources of knowledge, this would not entitle us to claim that only the scientific method gives us genuine knowledge of reality. Admittedly, it is true that if science does not go to its limits, it's a failure; but it is equally true that as soon as science oversteps its limits it becomes arrogant — a know-it-all, a case of

9   Gilbert Ryle, "The World of Science and the Everyday World," *Dilemmas* (Cambridge: Cambridge University Press, 1960), 68–81.

10   Larry Collins and Thomas Schneid, *Physical Hazards of the Workplace* (Boca Raton, FL: CRC Press, 2001), 107.

11   Edward Feser, *Scholastic Metaphysics: A Contemporary Introduction* (Neunkirchen-Seelscheid, Germany: Editions Scholasticae, 2014).

12   A. H. Maslow, *The Psychology of Science* (New York: HarperCollins, 1966), 15.

gross megalomania. No wonder this has led some to criticize scientism as a form of circular reasoning. The late philosopher Ralph Barton Perry expressed this as follows: "A certain type of method is accredited by its applicability to a certain type of fact; and this type of fact, in turn, is accredited by its lending itself to a certain type of method."[13] That's how round and round we go.

A fourth argument against scientism is that science is about material things, yet it requires immaterial things such as logic and mathematics. Logic and mathematics are not physical, and therefore not testable by the natural sciences—and yet they cannot be ignored or denied by science. Mathematical knowledge, for instance, is the most secure form of knowledge, but it is not about anything material. Other kinds of knowledge may arguably be more significant but that makes them harder to confirm. Einstein said it well: "As far as the laws of mathematics refer to reality, they are not certain; and as far as they are certain, they do not refer to reality."[14] Logic and reason are perfect examples of the kinds of immaterial phenomena that we all know exist, but that naturalistic science cannot measure, analyze, or account for. Yet these immaterial things are real and indispensable even though they are beyond scientific observation.

Take for instance the mathematical concept of π (pi). As Stephen Barr points out, it is not some private experience, like a toothache; it is not a material object like a melon; it is more than a sensation or a neurological artifact; it is certainly more than a certain pattern of neurons firing in the brain; it is not even a property of material things, for there are no pi-sided melons—perhaps close, but never exactly.[15] In other words, no material thing has the perfection that mathematical or geometrical entities have, so these truths do not depend on the material world. Instead, π is a precise and definite concept with logical relationships to other equally precise concepts. And concepts are bound to be mental, immaterial entities. To reduce them to a "creation of neurons" obscures the fact that "neuron" itself is an abstract concept. Those who claim that mental concepts are merely products of neurons should realize that talking about neurons requires the immaterial concept of *neuron* to begin with.

---

13 Ralph Barton Perry, *Present Philosophical Tendencies* (New York: Longmans, Green, and Co., 1912).

14 Address to Prussian Academy of Sciences, 1921.

15 Stephen Barr, *Modern Physics and Ancient Faith*, 194.

A fifth reason for rejecting scientism is that no science, not even physics, is able to declare itself a superior form of knowledge. Some scientists have argued, for example, that physics always has the last word in observation, for the observers themselves are physical. But why not say then that psychology always has the last word, because these observers are interesting psychological objects as well? Neither statement makes sense; observers are neither physical nor psychological, but they can indeed be looked at and studied from a physical, biological, psychological, or statistical viewpoint—which is an entirely different matter.

Often scientism results from hyper-specialized training coupled with a lack of exposure to other disciplines and methods. This may lead one to overlook the fact that the findings of science are always partial and fragmentary. In other words, there is no science of "all there is." There may someday be a "Grand Unified Theory" (GUT) in physics—a theory that unifies the three non-gravitational forces—but that is not the same as a "Grand Unified Theory of Everything." A theory of *everything* would also have to explain why some people believe that theory and some do not. Limiting oneself exclusively to a particular perspective such as physics is in itself at best a methodological decision, which some like to turn into a metaphysical decision. However, to quote Shakespeare, "There are more things in heaven and earth, Horatio, than are dreamt of in your philosophy."[16] One cannot give science the metaphysical power it does not possess.

A sixth argument against scientism is of an historical nature. The first legendary pioneers of science in England were very much aware of the fact that there is more to life than science. When the *Royal Society of London* was founded in 1660, its members explicitly demarcated their area of investigation and realized very clearly that they were going to leave many other domains untouched. In its charter, King Charles II assigned to the fellows of the Society the privilege of enjoying intelligence and knowledge, but with the following important stipulation: "provided in matters of things philosophical, mathematical, and mechanical."[17] ("Philosophical"

---

16   Hamlet, 1.5.167–68.

17   The Royal Society originated on November 28, 1660, when twelve men met to set up "a Colledge for the promoting of Physico-Mathematicall Experimentall Learning." Robert Hooke's draft of its statutes reads literally: "The Business and Design of the Royal Society is: To improve the knowledge of natural things, and all useful Arts, Manufactures,

meant "scientific" back then.) That's how the domains of knowledge were separated; it was this "partition" that led to a division of labor between the sciences and other fields of human interest.

By accepting this separation, science bought its own territory, but necessarily at the expense of all-inclusiveness; the rest of the "estate" was reserved for others to manage. On the one hand it gave to scientists all that could "methodically" be solved by dissecting, counting, and measuring; on the other hand these scientists agreed to keep their hands off all other domains — education, legislation, justice, ethics, and certainly religion — because those would require a different "expertise."

In spite of all the above objections, scientism is still very much alive, albeit mostly hidden "underground." The late Dutch physicist Hendrik Casimir (the Casimir effect of quantum-mechanical attraction was named after him) once said, "We have made science our God."[18] Indeed, science has become a semi-religion, of which scientists are the new "priests." Science is supposed to explain *everything*, but then in a much better way than God once did, in their view. It is in this frame of mind that Stephen Hawking once exclaimed, "[O]ur goal is a complete understanding of the events around us and of our own existence."[19] Scientism likes to broadcast far and wide, "It's all about science." Well, science may be everywhere, but science is certainly not all there is.

On the other hand, as we found out in the previous chapters, even an immaterial and immortal soul would need a material body with the "right" features. The immaterial faculties of language, rationality, morality, self-awareness, and religion do require a body that is capable of performing mental activities — which is arguably the outcome of a long evolutionary process. A gradual biological process was somehow preparing the way for the appearance of humanity. But denying that there is more to humanity than its biology is a form of scientism. If the soul is not material, then obviously it cannot be generated by merely physical processes such as evolution and genetics. There does not seem to be any evidence that a material system can generate the immaterial faculties of language, rationality, morality, self-awareness, and religion.

Mechanik practices, Engyries and Inventions by Experiments — (not meddling with Divinity, Metaphysics, Moralls, Politicks, Grammar, Rhetorik, or Logic)."

18 Quoted in: A. Van Den Beukel, *The Physicists and God*, 30.
19 Stephen Hawking, *A Brief History of Time* (New York: Bantam Books, 1998), 186.

Seen in this light, humans end up being radically different from their "relatives" in the animal world. Although we are flesh as they are flesh, we are also very exceptional creatures in this Universe, endowed with the faculties of language, rationality, morality, self-awareness, and religion. Genes do not determine which particular language we speak, which particular morals we have, and which particular religion we follow. But in addition, they may not even determine that we have the capacity to speak a language, to have a moral code, or to follow a religion.

Human beings have got to be more than what science tells us they are, otherwise science would in fact not have much to tell us. In addition to our five material senses we have at least three immaterial "senses" — the rational sense of "true or false," the moral sense of "right or wrong," and the religious sense of "material or spiritual." The origin of these three immaterial senses seems to keep eluding us.

INDEX

ABOUT THE AUTHOR

GERARD M. VERSCHUUREN is a human geneticist with a doctorate in the philosophy of science. He has studied and worked at universities in Europe and the United States. Currently semi-retired, he spends most of his time as a writer, speaker, and consultant on the interface of science and religion, faith and reason. His most recent books from Angelico Press are: *Life's Journey — A Guide from Conception to Growing Up, Growing Old, and Natural Death*; *Aquinas and Modern Science — A New Synthesis of Faith and Reason*; and *The Myth of an Anti-Science Church — Galileo, Darwin, Teilhard, Hawking, Dawkins*.

www.ingramcontent.com/pod-product-compliance
Lightning Source LLC
Chambersburg PA
CBHW021359090426
42742CB00009B/922